Watermills & Windmills
of Middlesex
has been published
as a Limited Edition
of which this is

Number

Edmonton post mill in its prime, from a watercolour of 1890. (FHM)

WATERMILLS & WINDMILLS of MIDDLESEX

BY

GUY BLYTHMAN BA

BARON
MCMCVI

PUBLISHED BY BARON BIRCH
FOR QUOTES LIMITED IN 1996
AND PRODUCED BY KEY COMPOSITION,
SOUTH MIDLANDS LITHOPLATES, CHENEY & SONS,
HILLMAN PRINTERS (FROME) AND
WBC BOOKBINDERS MANUFACTURERS

ISBN 0 86023 538 6

Contents

ACKNOWLEDGEMENTS

A great many people have helped in various ways with the research for this book, and I would like to record my thanks to them for their assistance. They are: Mr D. Austin (Barnet Musuem); Mr T. F. T. Baker (County Editor, *Victoria County History*); Ms J. Banjam (Leighton House Museum); Ms M. Barber (Lambeth Palace Library); Barnet Archives & Local Studies Centre; Mr D. N. Barriskill, Mr S. French and staff at the Guildhall Library; Mrs E. M. Bayliss (Fulham Library); Clive Birch, my publisher; Mrs J. Bond; Mr R. Bowden and staff at the Marylebone Library; Mr M. Brant; Mr B. Brennan; British Museum and Library; Paul Brown, who kindly agreed to do many of the line drawings; Bruce Castle Museum; Mrs Ella Bubb for translating documents in the Greater London Record Office and Guildhall Library; Mr Stephen Buckland for permission to reproduce his line drawings and for general assistance; Mr W. Burnett (Sunbury windmills); Miss A. Cameron (Treaty Centre Library, Hounslow); Dr T. H. Cashmore for information on windmills and watermills in the Twickenham area; Mrs M. Chamberlain; Mr Peter Chaplin (for information on Sunbury windmills); Chelsea Library; Mrs D. M. Clarke for translating documents in the Public Record Office; Ms P. Clarke (Pinner Local History Society); Mrs C. Cotton (Uxbridge Library); Professor Alan Crocker; Mr Graham Dalling (Local History Officer, Enfield Public Libraries); Mr N. J. Davies of the Greater London Site & Monuments Record, Museum of London; Mr Peter Dolman; Mr C. Eve; Mr B. Firth (Greater London Industrial Archaeology Society); Dr Lynne Fox; Mr V. George; Mr M. Gibson (Central Library, Islington); Mr J. Griffin (Forty Hall Museum) for permission to view illustrative material in the museum's collection and to reproduce the watercolour painting of Edmonton mill; Ms V. Hart (Central Library, Islington); Mrs M. Gooding of the Local History Library, Ealing; The Greater London Record Office and History Library; Mr Roy Gregory; Mr P. Hammill; Mrs Carolyn Hammond (Chiswick Library) for information regarding windmills and watermills at Brentford; Mrs J. Harding (Ealing Museum, Art and History Society); Mrs A. Harper-Smith (Acton History Group); Dr T. Harper-Smith (London & Middlesex Archaeology Society); Mr Julian Harrap and Ms Caroline Wilson of Julian Harrap Architects; Mr B. A. Harrison (Pinner Local History Society); Mr Dennis Haselgrove (Fulham Local History Society); Mr R. Hawksley; Mr K. Y. Heselton; Mr H. M. Hoather; Mr M. Holmes and staff at the Swiss Cottage Library, Camden; Ms D. Howard; Mrs P. Ivie; Mr G. Javes (Barnet Museum); Ms Bridget Keane (Grange Museum, Neasden); Ms C. Kelter; Ms S. Kirby; Mr R. Knight (Local Studies Centre, Swiss Cottage Library); Mr R. S. Knight (Borough of Twickenham Local History Society); Mr G. Lagrue; Dr John Langdon for permission to use material from his research into medieval windmills and watermills; Mrs E. Leeson; Mrs J. Lewing (Gunnersbury Park Museum); Mrs M. H. Lodge (Hounslow and District Local History Society); London Borough of Tower Hamlets Local History Library and Archives; Ms Jean Macdonald (London and Middlesex Archaeology Society); Mr J. K. Major; Ms H. Maltby (Central Library, Royal Borough of Kensington and Chelsea); Mr D. Mander and staff at the Hackney Archives Department, Rose Lipman Library, Hackney; Mr & Mrs B. Marsden; Mr J. Marshall (Treaty Centre Library, Hounslow); Mr N. Moon; Mr R. Morgan (Wembley History Society); Mr R. Mortimer (Keeper of Muniments, Westminster Abbey); Rev D. J. Nash; The

Newcomen Society for permission to reproduce illustrations and text from Stephen Buckland's article on 16th and 17th century London windmills in the 1989 edition of their *Transactions*; Mr M. Newland-Smith of the River Lea Tidal Mills Trust; Mr P. A. Nicholas; Miss L. Noble (Willesden Local History Society); Mr C. Nugent of the Wentworth Art Gallery, University of Manchester, for arranging for a print to be made of the 1825 painting of Acton smock mill, and for permission to reproduce same; Mr L. O'Flanagan; Mr David Pam; Mr Y. Parfitt; Mr Kenneth Prater; Mr C. Pratt; Dr G. J. Pritchard; Ms Jane Ramsay (Camden History Society); Mrs J. Roylance-Court; Mrs B. Richardson; Mr E. L. Scott; Mr Colin Shrimpton; Miss J. T. Smith of the Essex Record Office for permission to view material in the papers of the late Kenneth G. Farries; Mr Roland W. Smith; Ms Caroline Starren and staff at the Local History Library, Chelsea; Mr J. Stevens (Enfield Archaeological Society); Surrey County Record Office; Ms M. Swarbrick and staff at the Victoria Library; Miss M. Swatton (West Drayton & District Local History Society); Thames Water for permission to inspect the remains of the windpump at New River Head, Clerkenwell; Dr K. Thompson; Mr R. Thomson (Civic Centre Library, Harrow); Mrs Nan Trimble; Twickenham Public Library; Mr K. J. Walker; Ms Joan M. Warren; Mrs E. Watling (Ruislip, Northwood and Eastcote Local History Society); Mr B. T. White (Hayes & Harlington Local History Society); Mr F. Willison; Mrs G. Wood; Mr D. Woodyard; Mr E. Wright; Ms M. Wright; Mr David Wright of G. R. Wright & Sons; Mr B. Wrigley (Hendon and District Archaeological Society); Mr J. Young (Pinner Local History Society); various residents of Sunbury and Shepperton, namely Mrs Alexander, Mr Coltman, Mr Connor, Mrs Heath and Mr Tulett, for information on windmills in the former place; and finally to my family, especially my mother, Mrs Diane Bythman, for their support and patience during times when the project rather came to dominate their lives. To anyone whom I may unwittingly have omitted from this list I offer my apologies.

Dedication

For my father, Michael Arthur Blythman, with whom I spent many happy hours looking at windmills in various parts of Britain and Europe — if only he could have lived to see the publication of my first windmill book.

The Watermill

Listen to the Water-Mill
Through the live-long day;
How the clicking of its wheel
Wears the hours away.
Languidly the Autumn wind
Stirs the greenwood leaves;
From the field the reapers sing,
Binding up their sheaves.
And a proverb haunts my mind
As a spell is cast;
"The mill cannot grind
With the water that is past."
Autumn winds revive no more
Leaves that once are shed,
And the sickle cannot reap
Corn once gathered;
And the ruffled stream flows on,
Tranquil, deep and still;
Never gliding back again
To the Water-Mill.
Truly speaks the proverb old,
With a meaning vast,
"The mill cannot grind
With the water that is past."
Take the lesson to thyself,
Loving heart and true.
Golden years are fleeting by,
Youth is passing too.
Learn to make the most of life,
Lose no happy day,
Time will never bring thee back
Chances swept away.
Leave no tender word unsaid,
Love while Love shall last;
"The mill cannot grind
With the water that is past."
Work! While yet the daylight shines
Man of strength and will;
Never does the streamlet glide
Useless to the mill.
Wait not till tomorrow's sun
Beams upon thy way;
All that thou can call thine own
Lies in thy today.
Power, and intellect and health,
May not always last,
"The mill cannot grind
With the water that is past."
Oh! The wasted hours of life
That have drifted by;
Oh! The good that might have been
Lost without a sigh.
Love that we might once have saved
By a single word;
Thoughts conceived but never penned,
Perishing unheard.
Take the proverb to thy heart,
Take and hold it fast;
"The mill cannot grind
With the water that is past."

Bishop Heber

INTRODUCTION

Originally this book was to be about windmills only. They have long held a certain fascination for me, and it seemed appropriate that I should one day write a book about them. Watermills seemed to have none of their distinctive character, and therefore attracted me somewhat less; I suppose I was rather prejudiced against them.

There having already been many windmill books of a general nature, I wished to tackle the mills of a particular region. Middlesex was chosen for two reasons. Firstly, no comprehensive account of its mills had previously been written. Some, in particular Mr Denis Sanders of Feltham and Mr Kenneth Farries, two molinologists (students of mills) of repute who are sadly no longer with us, had had plans for one, but these failed to reach fruition. Secondly, I could say it was my home county. I was born at Hampton and lived there until we moved to Twickenham in 1971. In 1976 we left for Frimley in Surrey, remaining there for eleven years before returning to Middlesex to live at Sunbury-on-Thames, where I have been ever since.

I has always been frustrated by the virtual absence of windmills in the area. The windmill enthusiast living in the former county of Middlesex suffers acutely from 'mill starvation', usually needing to travel quite a distance to find the nearest specimen. In other parts of London windmills can still be found, for example at Upminster (formerly Essex), Arkley (Hertfordshire), Keston (Kent), Brixton, Shirley, Wandsworth Common and Wimbledon (Surrey), but no windmill sails may be seen peeping above the rooftops of Hounslow, Twickenham, Ealing, Harrow, Hackney, Fulham, Hampstead, Hendon or Enfield. Just why this is so is difficult to say, for several of the mills survived into an age when the need to preserve industrial monuments was becoming accepted, but bad chance probably had something to do with it. Windmills belong to an essentially rural environment which in Middlesex disappeared so long ago that it is hard to believe any ever existed there. But exist they did, and the dearth of physical remains, which lent to the subject an air of mystery, only made me more determined to write a book on it.

While regrettable, the fact that most of the mills had disappeared well before the turn of the century made my task a lot easier. There was no fieldwork to be done, no fighting one's way through mud and clinging vegetation to photograph what might be litle more than a few crumbling bricks (for to the student of mills, whatever remains must be recorded). The collection of notes on British windmills and watermills compiled between 1932 and 1973 by H. E. S. Simmons, held at the Science Museum Library, and the *Victoria County History*, proved valuable starting points. Public libraries and museums in the former county, and the Greater London and Public Record Offices, provided a variety of sources: books, photographs and other illustrations, maps, newspaper cuttings, trade directories, and documents relating to ownership the main ones. The research took five years, from 1988 to 1993.

The decision to include watermills was taken comparatively late in the day. I felt it would add considerably to the book, especially as several examples still exist. They proved quite as fascinating to research as the windmills.

No doubt there is more information to be discovered. In the meantime I hope I have managed to produce something of interest to both the students of mills and the general reader, to Middlesexians and non-Middlesexians, and to shed light on an aspect of the county's history that tends to be forgotten.

ABOVE: House Mill, Bromley-by-Bow (383828), early this century. This fine large tide mill, built in 1776 and last worked in 1941, stood historically just outside Middlesex. It is currently being restored to working order by the River Lea Tidal Mills Trust. (RLTMT) BELOW: Clock Mill, Bromley-by-Bow, 1994 was also a tide mill, with two distinctive oast houses and a clock tower — hence its name. Now offices, the wheel and some of the machinery remain but have been bricked in. (GB)

WATERMILLS

The watermill has a much longer ancestry than the windmill, one which goes back to classical times. We cannot be sure whether it was the Greeks or the Romans who invented it, nor when exactly it replaced the cumbersome and inefficient practice of grinding corn manually by pounding it between two stones or using a hand quern.

Watermills were fairly widespread in Anglo-Saxon England, as the number recorded in the Domesday Book of 1086 indicates; a 'mill', as the people of the time classified it, could mean any set of machinery driven by a waterwheel, rather than the building in which it was situated, and many buildings had more than one within their walls, so the actual number of watermill structures was not as large as it might appear). According to Domesday there were 30 and a half mills in Middlesex, listed here with their worth: Stibenhede (Stepney), 4, amounting in total to £4 16s 4½d; Hesa (Hayes), 1: 4s; Draitone (Drayton), 1: 13s 5d; Staines, 6: 64s; Hanewell (Hanwell), 1: 2s 2d; Covelie (Cowley), 1: 5s; Hermondeswerde (Harmondsworth), 3: 60s, 50 eels; Coleham (Colham), 2: 41s. Also half a mill (see explanation of this in entry on Colham in watermill gazetteer) worth 5s; Adelmetone (Edmonton), 1; 10s; Enefelde (Enfield), 1: 10s; Chingesberie (Kingsbury), 1: 3s; Stanwelle (Stanwell), 4: 70s; Gistelesworde (Isleworth), 2: 10s; Herefelle (Harefield), 2: 15s.

As with windmills, the number of watermills increased with population growth. In Middlesex, although the availability of water was a factor limiting the spread of the watermill, from which of course windmills did not suffer, the existence of the Thames, along with other rivers and their tributaries, was an obvious spur to development. Initially watermills, like windmills, were owned by the lord of the manor, one of whose tenants was miller. In Middlesex this often remained the case until well into the 19th century, when wealthy landowners like the Duke of Northumberland still owned large areas. In mediaeval times the tenants were obliged to grind their corn at the lord's mill — he took a certain share of the profits — and nowhere else, facing a severe fine or worse if they failed to comply.

The greatest concentration of mills was in the Uxbridge area, which appears to have been the industrial equivalent in Middlesex of Lancashire and the West Riding of Yorkshire. They tended to have as grim and forbidding an appearance as their counterparts, whether driven by water or steam, in those northern regions. They owed their spread to the Grand Junction Canal which, along with a high level of agricultural productivity, made the area an important centre of flour milling. During 1799, 4,612½ tons of flour were transported along the Canal. The millers of Uxbridge were characterised by dissent and Quakerism.

Many sites were occupied by a mill for 800 years or more but, as with windmills, Saxon or mediaeval structures and those which survived into the 19th or 20th century were rarely the same buildings. Mills would be periodically rebuilt to take advantage of technological improvements, enlarged to accommodate additional millstones and machinery, so that the site plans changed considerably. A large timber structure might be rebuilt in brick, like Clock Mill at Bromley-by-Bow just over the Essex border, although it was common for a timber-built mill to remain so throughout its working life, the Ponders End mill at Enfield providing one example. The final stage in the development of the watermill was represented in Middlesex by the huge brick-built Kidds' Mill at Isleworth with its 29 pairs of stones.

The earliest watermills, small wooden constructions, had the wheel fixed horizontally to the lower end of a vertical shaft on whose other end was mounted the upper of a single pair of stones. (The upper or runner stone rotated while the lower or bedstone was always stationary, this remaining the case right through to the eventual demise of stone grinding). Later the wheel was mounted vertically on a horizontal shaft, at the other end of which was another vertical wheel usually situated within a kind of pit and therefore called a pit wheel. The cogs or teeth of this wheel engaged with those of a horizontal wheel known as a wallower, the latter being carried on an upright shaft. Also on this shaft, above the wallower, was a great spur wheel which usually drove the runner stones from underneath *via* small gears called stone nuts on shafts or quants, the latter passing up through the centre of the bedstone and being fixed to that of the runner stone. A crown wheel further up drove hoisting equipment and additional machinery (*ie* for extracting the flour from the ground corn, or meal, by breaking it down into its various components, a process known as 'dressing'), *via* belts and pulleys, bevel gearing or both. In only a few cases was the great spur wheel above the stones. Windmills were 'overdriven' or 'underdriven' in roughly equal amounts.

Wheels came in several different types, according to the point at which the water hit them (they were usually constructed with two sets of spokes and two rims between which were fixed buckets, the weight of the water in these, plus the force with which it struck the wheel, being what made them turn). If it was directed onto the top of the wheel, the latter was called an overshot wheel; if it struck the wheel more or less halfway up, a breast shot; if impacting on the lower part, an undershot. The pit wheel and associated machinery were usually located on the lowermost floor; above this was the stone floor which housed the millstones. The number of storeys in a watermill varied with size, the largest having five or more. The uppermost storey, within the roof, normally contained the bins for storing grain, on either side of a central walkway.

The wheel was by no means always driven directly by the river or stream; sometimes it was channelled off by means of a cut. At some mills it was stored in a mill pond and directed to the wheel by another cut. The amount of water reaching the wheel could be controlled by a sluice gate.

The need for millers to draw off a certain amount of water from the river for their mills sometimes brought them into conflict with others who used it, whether for commercial or recreational purposes.

Whereas the windmills were rarely used for anything other than grinding corn (their second most common purpose being drainage, particularly in parts of East Anglia), instances of watermills being employed for other 'industrial' use were more common. In Middlesex we know of silk, leather, gunpowder, brass, mustard, paper and millboard (a form of cardboard often used in books) mills. The structure of a windmill, particularly in the case of post mills, appears to have been less suited to the kind of machinery needed for these purposes.

Watermills were just as affected by progress as windmills, but not necessarily in the same ways, and over a much longer period. Since they were not dependent on the wind, of which a mill could be robbed by too much building in its vicinity, they were not as adversely affected by urban development. Watermill buildings were also more easily adaptable, or susceptible to alterations and improvements, unlike the structure of a windmill with its own special requirements and restrictions. Windmills tended on the whole to be smaller and narrower, with less room for new plant. It is principally for these reasons that many watermills remained in use until well into the present century while their wind-powered cousins, particularly in Middlesex, tended to cease operating earlier. Modifications, such as the replacement of waterwheels by turbines and of stones by roller plants, ensured that the mills continued to operate efficiently and thus remain economical. Some survived to become power mills, driven by steam, gas, or electricity. Even when a mill had altogether ceased as such, the building might in some way contribute to a business, as has happened at Ponders End, where it now serves as a warehouse. The Upper Mill at Stanwell houses a chemical company, while Clock Mill, just outside Middlesex, has been converted to offices. The remains of Black Jack's Mill at Harefield have been incorporated into a private house while Fountains Mill has served the local community in several different ways.

It is principally for these reasons that half-a-dozen watermills are still standing in the county today, whereas the windmills have virtually all disappeared, although none are in anything like a complete state. Watermills may not have been altogether as common or widespread as the windmills but in this respect at least they were undoubtedly more successful.

The examples remaining today, all more or less devoid of machinery, are Stanwell (Upper Mill), Enfield (Ponders End), Uxbridge (Fountains Mill), Cowley, Yiewsley, and Harefield (Black Jack's and the copper mill). There are also some remains at West Drayton and Bedfont (the gunpowder mills).

House Mill, Clock Mill and the windmill which stood near the watermills, also on the Essex side of the Lea; watercolour by Samuel Beker, 1746. The Clock Mill was then made of timber. The windmill has common sails on which the canvas is unfurled, at work when the picture was painted. The decorative finial on the roof is typical of many post mills built in the seventeenth century and the first half of the eighteenth. (MOL)

ABOVE: House Mill, 1994, (GB) with LEFT: part of iron waterwheel with section of broken rim (GB) and RIGHT: iron waterwheel with wooden buckets. (GB)

LEFT: House Mill: control gear for the sluice gates; (GB) RIGHT: one pair of millstones, still with its wooden casing, or tun, and feeding mechanism intact. The grain is fed into the hopper which is mounted on a wooden framework called a horse. From this it flows down an inclined trough or shoe into the hole or eye in the centre of the runner stone and passes between it and the bedstone to be ground; (GB) BELOW: a stone with its runner removed, showing the eye and a lead weight inserted to keep the stone level. The weight is by Corcorans of Mark Lane, London; (GB) RIGHT: layshaft and pulleys from which ancillary equipment, now removed, was driven. (GB)

ABOVE: House Mill: the sackhoist drive, (GB) and BELOW: view of bin floor. (GB)

WATERMILL GAZETTEER

ACTON

(1-3) Early Mills, sites unidentified.
(4) Steyne Mills, Stamford Brook. Standing 1864. 199805

There were four watermills at Acton Manor in the sixteenth century, according to the *Victoria County History*, although the site of only one has been located with certainty. A place known as Steyne Mills included a 'mill house' in 1722, although it was then a tanyard, and a watermill is shown there on the 1864 Ordnance Survey.

BEDFONT

(1) Lower Mills 112746 — Duke of Northumberland's River.
(2) Upper Mills, 108746 — Duke of Northumberland's River.

The gunpowder mills at Bedfont, like those of Hounslow, have attracted a great deal of interest from industrial archaeologists and have been well chronicled elsewhere.

In 1609 King James I granted permission to build powder mills in the area. They are thought to have been a reconstruction of a corn mill, one of two mills which stood on the Duke of Northumberland's River between Faggs Road and the River Crane, on land to the northwest of Baber Bridge. The other was a paper mill and under separate ownership. The venture was short-lived; by 1630 the powder mills had been converted into a sword mill, where sword blades, used by both sides in the Civil War, were manufactured. After the mid-1650s, production ceased and in 1655 Thomas Carter, who had acquired the lease, converted the mills back to the manufacture of gunpowder. Twelve years later he established a second powder mill to the south of Baber Bridge, and a later owner of the sites, John Richardson, added the paper mill to the complex, converting it too into a gunpowder mill. The original gunpowder mills now became known as the Upper Mills, and the former paper mills as the Lower Mills. The Upper Mills ceased production as powder mills in about 1750 but were converted back to the making of explosives in the late 19th century as the North Feltham Cartridge Factory. Today, nothing remains of them apart from the mill race.

In 1799 the Lower Mills were destroyed by an explosion and subsequently rebuilt on what was considered a safer plan, spaced further apart and further from the road. Mounds of earth called 'baffle mounds', intended to contain and reduce the force of any explosions, were thrown up near the buildings. The complex was smaller that that at Twickenham (the Hounslow Gunpowder Mills), from which it was originally independent but whose owners later acquired it as an auxiliary site. The safety record of the Bedfont mills was better than that of Hounslow, but they had their fair share of accidents — in 1795 two men were killed, and again in 1825. The inevitable depression in the gunpowder industry after the First World War caused the closure of both works during the 1920s, the buildings being subsequently dismantled and set alight to destroy any residual gunpowder.

Some remains of the buildings may still be seen amid pleasantly wooded surroundings, and a visit to the site is well worthwhile. A guide to the site and the walks which may be had there can be consulted at the Treaty Centre Library, Hounslow.

The mills were built largely of wood so that they could be more easily and cheaply reconstructed in the event of an explosion. They stood on brick bases, which constitute the main visible remains. They stand on either side of the river, which was divided by a brick pier into two mill races, so that it drove the wheels of both mills. This pier also provided support for the ends of the axles carrying the wheels. The bricked-up circles where these entered the walls of the mills are still visible. The machinery within was rather different in arrangement from that of a corn mill: the great spur wheel engaged with another at the top of a shaft to whose lower end was fixed a pair of vertically-mounted millstones — known as edge runner stones — which revolved above a bedstone. The ingredients of gunpowder, namely saltpetre, sulphur and charcoal, were placed on the bedstone and water added. The runner stones would then revolve over the mixture to produce a kind of cake. In the 1890s the mills were modernised with the introduction of steam engines, and the boiler house and supporting brickwork is still evident, with the threaded rods for bolting the engines into place. Apart from these remnants little is left of the complex apart from the baffle mounds and the occasional millstone.

BOW

River Lea. Paper mill, Bow Bridge. 379834

Edward Lloyd, owner of the paper mills at Sittingbourne in Kent, which remain functional to this day, also operated one at Bow Bridge during the 1860s and 1870s. Lloyd was the proprietor of a weekly newspaper which bore his name, and it was for this that the paper was intended. They were in operation in 1862, when Lloyd was granted provisional protection for improvements in the machinery used in the process. They were then turning out straw and rag printings, but by 1868 produced news and rag printings from raw fibres. Unusually, the mill never appears to have received the Excise number normally allocated to paper mills.

The mill evidently ceased operations during 1878; it is listed in a *Directory of Paper Mills* for that year, but in another (*Kent's*) published on 1 January the following year, it is included in a list of mills struck off the register, being described as 'broken up'.

BRENTFORD

(1) Early mill.
(2) Old Brentford. Standing 1564.
(3) River Brent, at The Buttts. 175776
(4) River Brent, at Brentford End. Gone during 19th century. 175774
Plus various mills whose sites cannot at present be identified.

In 1297 the lord of the manor of Isleworth owned a mill 'of Brentford'. A watermill was built over a common sewer, whose water provided the power for the wheel, at Old Brentford some time prior to 1564. In 1738 the lord of the manor leased the right to build a mill on the Brent; this stood at the western end of the district known as the Butts until its demolition in 1904, by which time it was showing signs of dilapidation. It was a small and undistinguished-looking building, part brick and part timber with a tiled roof. Comparatively little is known about this mill's history. After ceasing work it was home to a firm of builders and decorators. Following this, the Canal Boat Mission negotiated its purchase from the owners, the Grand Junction Canal Company, but it was decided instead to pull the the existing structure down and build a new one.

Brentford End
This mill stood on the Brent just above Brentford Bridge. In 1768 it was owned by Richard Dixon and William Allcock, of Cripplegate and Covent Garden reespectively, and tenanted by Edmund Hill. It was then a timber affair with a tiled roof. By 1777 it had been partly rebuilt in brick; Hill was still tenant but the owners were now Thomas Harrington and Edward Kendall. Two years later Hill owned the property. An insurance policy note of 1786 indicates the mill was by then entirely of brick, adding it is 'well secured by an Iron Door'. Thomas Hall was the tenant miller. By 1793 he had been succeeded by Messrs Hill and Stanborough, who operated the mill until 1837 when their partnership was dissolved. The mill appears to have gone by 1865 — it is not shown in the Isleworth Inclosure Award.

In the 18th and 19th centuries Brentford was an important centre of the milling industry, and in 1792 there are said to have been five watermills there altogether, including those at Brentford End and the Butts. The following insurance policy notes all relate to watermills, but it is impossible to say where they stood:

(1) A brick- and tiled mill was insured against fire in 1782 by Thomas Finch of New Brentford for £400.

(2) In 1788 Richard Bax had a watermill at Brentford, within which he kept stock to the value of £300.

(3) In 1793 Dr Robert Wallace Johnson and Blissett William Gould had a mill, again brick- and-tiled, at the wharf, which they insured for £300. Here water power was supplemented by a steam engine.

CHISWICK

River Thames. 15th century mill.

Sutton Court Manor had a watermill 'next Sutton', in the area of the present Sutton Court Road, in 1458; it had gone by 1590.

CITY OF LONDON

(1) River Thames. St Katherine's Dock 345805 (approx.)
(2) River Thames. Traitor's Gate 817805
(3) River Thames. Crash Mills, Wapping. 343803

St Katherine's Hospital near the Tower of London, founded in 1148, had a watermill on the site later occupied by the Iron Gate. William de Longchamp, Richard I's Chancellor during his absence on the Crusades, found that the mill stood in the way of his planned alterations to the Tower and had it demolished, to the chagrin of the Hospital. Eventually, as part of the reconstruction of the Tower during the reign of Edward I (1271-1307), a new mill was provided as compensation. It was standing in 1580 when it was described as 'the Abbot's Mill of Tourehill'. The site of the Hospital is today covered by St Katherine's Dock.

In the 18th century there was a waterwheel for pumping water at Traitor's Gate. Both this and the above mill derived their power from water pent-up in the Tower moat by high tide.

A tidal watermill also stood at the intersection of Wapping High Street and Thomas More Street. The 'Mill of Crosseine Lane', as it was called in 1235, originally belonged to Holy Trinity Priory, Aldgate. Later it became the property of the Abbey of St Mary Graces on Tower Hill, founded by Edward III (r.1327-77). In 1603 certain people were charged with obstructing by means of 'Clay, posts and mud' a watercourse which linked the Thames to 'divers mills called Crashmilles'. The mills were last heard of in 1608 when the Crown, which by now owned them, was leasing then at £16 per year. Like the Hospital mill, all traces have been obliterated by dockyard construction.

CLERKENWELL

River Fleet. Various mills from the 14th to the 18th century.

Here, the lower reaches of the River Fleet were known in the 14th century and after as Trulmyll, Turnmill or Tru-mill Brook because of the number of watermills on it north of the Fleet bridge. The noise they collectively produced was considered pleasant by the mediaeval chronicler Mark FitzStephen, who described the area as 'a delightful plain of meadow land interspersed with flowing streams, on which stand mills, whose clack is very pleasing to the ear'. This testifies that mediaeval technology could actually be an environmental asset, in sharp contrast to that of later centuries. The Turnmill Brook gave its name to Turnmill Street, a name which still survives. In FitzStephen's time two of the watermills were owned by the Knights of St John of Jerusalem and let by them to one Francis Bach at a rent of 100s per year. Another belonged to the nunnery of St Mary, Clerkenwell. A description of the land owned by the nuns in the register or Cartulary of the nunnery mentions 'the ditch that goes from Holeburne to the mill of the nuns'.

It stood close to the gardens and fishponds of the Knights Hospitaller, who owned the next mill along; this had two waterwheels.

In the time of King John, there was a mill near Castle Lane, between Blackfriars and the Thames, belonging to the Knights Templar. John granted a place in the Fleet River, near Baynard's Castle, to build a tide mill, and the whole course of the water of the Fleet to serve the mill.

In 1307 Henry Lacy, Earl of Lincoln, complained that by, among other factors, 'the diversion of the river by them of the New Temple for their mills without Baynard's Castle', ships could not navigate the river where it ran under Holebourne Bridge and Fleet Bridge into the Thames. On the petition of the Earl, the Constable of the Tower along with the Mayor and the Sheriffs of London were directed to procure certain 'honest and discreet men to inquire into the former state of the river', and return it to its original course and condition. The mills were removed, and other steps taken , but they seem not have been effective. The river was not restored to its proper course or breadth, and in time came to be regarded as merely a brook. Mills again began to make their appearance on it; however, 'it was soon choked with filth again'.

Pollution was frequently a problem. The river had to be cleaned out in 1502 after which it continued to drive several mills, for corn grinding and other purposes, but later returned once more to an unsavoury state. The last time the Fleet was capable of turning a mill-wheel was in the 18th century. In 1714 there appeared in the *Daily Courant* of 17 September an advertisement for a house to let in Bowling-Alley, Turnmill Street, with the advantage of a 'common sewer, with a good stream and good current' running by it 'that will turn a mill to grind . . . powder liquorish and other things'. A snuff mill also stood there, and in 1740 a printer named Cave purchased a machine for spinning wool or cotton into thread, yarn or worsted, consisting of 100 spindles, and built a watermill on the Brook to drive it. Its upkeep was the responsibility of Paul of Birmingham, who had originally patented it. However, the mill was never brought into profitable order, and shortly after this smells had become such a nuisance that most of the river was paved over to form a road.

COLHAM/HILLINGDON

(1-3) Domesday Mills (see above).
(4) River Colne. Port Mill, later Lower Colham Mill. 057801

In Domesday Book, there were two and a half mills in Colham manor; presumably the 'half' represents a mill driven by the main stream of the Colne, and therefore probably belonging in part to a manor on the Bucks bank. During the Middle Ages two mills belonging to Colham, presumably on the sites of the Domesday mills, were called Port Mill and Bury Mill. Later they became respectively Colham or Lower Colham Mill and Yiewsley Mill.

The former was separated from the Manor in 1771 and sold to John Hubbard, a mealman, members of whose family had tenanted it since at least 1690. Hubbard leased it within the next year or so to James Barrett & Co. In 1788, when insured against fire, the mill was worth £600.

A watercolour of c1805 depicts the mill as a timber building on a brick lower storey with a tiled roof in which two dormer windows are visible, as well as a couple of dovecote-like structures which may have been for ventilation. In 1800 it was owned by John Hubbard's son, also John, Rector of Shepperton, but by 1832 it had been acquired by Thomas Smith, working in partnership with his son Joseph as mealmen and millers. They ceased to trade as mealmen in 1839, but continued as millers until that partnership too was dissolved at the end of 1843. After this Joseph Smith ran the business on his own. By 1862 he had been succeeded by Edward & John Fountain, later of Fountains Mill in Uxbridge, the mill itself becoming known as 'Fountains Mill' as did the Uxbridge mill in its turn. They operated up to 1895. In 1891 they offered the mill for sale; an advertisement in a local paper projects two waterwheels driving eight pairs of stones. Attention was drawn to the mill's advantageous position close to West Drayton station. The sale was unsuccessful and the mill was readvertised the following year, pointing out that it could turn out around 500 sacks of flour per week, and that this could be increased.

The mill was later acquired by the Rotary Photographic Works. By the time it ceased work the largely timber building had been rebuilt in brick. No trace remains.

COWLEY

(1) Domesday Mill (see above).
(2) Fray's River. Rabb's Mill. 054933
(3) Fray's River. Cowley Hall Mill. 048818

Rabb's Mill, which belonged to Cowley Hall, was a comparatively recent affair, first recorded in 1816. The miller in 1823-4 and 1832-4 was John Austin of Hillingdon, later succeeded by E. Fountain & Son, also of Uxbridge and Colham mills, who were there in 1851 and 1855. By 1878 it had been acquired by Grimsdale & Sons, corn merchants of Uxbridge, who also traded under the name of Dobell & Co.

It was totally destroyed by fire in 1898, shortly after being refurbished with new machinery and a powerful gas engine, but was subsequently rebuilt. The fire was a mystery as the mill was not in operation. The new mill ran until March 1928 when it too was ravaged by fire. The heat was so intense that one fire engine remained on the scene for two days, while the ruins smouldered for over a week. The insurance company was not satisfied and this time the mill was not rebuilt, Dobells subsequently being wound up. The remaining walls were demolished in the late 1930s and a laundry built on the site; at the same time the waterwheel, which had survived the blaze, was removed. The mill race was still to be seen in the early 1970s.

A photograph shows the mill as a rather grim Victorian affair, brick-built with a tile mansard roof, on which stands a square wooden structure whose purpose is unknown. An inscription on the iron sluice gate declared it to be the work of Joseph J. Armitage of Ringwood, Hampshire, whose engineering firm specialised in manufacturing equipment associated with watermilling.

A new Cowley Hall mill was built to replace that burnt in 1864. It had two storeys, the lower of brick and the upper, which once housed a joiner's shop, largely of wood. The upper section is of special interest in that one one side the weatherboarding is laid vertically and is inside the main framing. To leave the timbers exposed to the elements in this fashion was rare, and not perhaps sensible, practice, The building, which is in some need of repair, contains two turbines which are under a permanent head of water. Latterly the mill was used to generate electricity for the nearby house, and some of the equipment for this is still in place. The fine house attached to the mill is now ruined.

CRANFORD

In 1650 the inhabitants of Cranford, along with those of the neighbouring parishes, petitioned unsuccessfully for a corn mill to be built on the Duke of Northumberland's River near the sword-blade mills on Hounslow Heath.

WEST DRAYTON

River Colne. 054791

This is one of those places where a mill has stood since mediaeval times. It appears to have always occupied the same site, and no other is known to have existed in the parish. After its mention in Domesday it is recorded in a survey of the manor in 1222. In 1467 the Dean of St Paul's leased it for 99 years to Clement Cook of Twickenham. After passing through various hands the lease was bought by William Hyll, who assigned it in 1538 to William Paget, whose property it became, along with the manor, in 1546 when the Dean and Chapter of St Paul's surrendered their holdings to the Crown. In a survey of the manor made by Edmund Twynyho, Lord Paget's auditor, in November 1557 reference was made to the 'new mylle' and the 'old mylle'; it had two mills, for corn and malt, under one roof. The value of the mill together with the fishery, watercourses and weir, one acre of land called Chavingcrofte, a rood of land called an Aldergrove, and two 'eights' (aits, which are small islands) was £18; the miller, John Baker, was required to make this payment in two parts, at the feast of the Annunciation (25 March) and at Michaelmas (29 September). Paget had enlarged the mill by 1559, adding an additional wheat mill. Under the leases the miller was responsible for grinding the lord's corn free of charge and for keeping the mill stream clear of rubbish and its banks in good repair.

In 1696 the mill was leased to Nicholas Faulcon, who used part of the premises for paper-making. Repairs were carried out in 1697-8. In 1736 James Mills secured a lease, which passed on his death in 1764 to his wife Elizabeth, who continued the business with their son, Alan. Elizabeth bought it in 1772, and discontinued the grinding of corn altogether. Shortly before 1796 the mill was in the possession of Nicholas Mercer, who almost wholly rebuilt it. His family was originally from Uxbridge, but soon established themselves in Drayton. They grew to have so many local connections that by the 1880s it was said to be impossible to walk down the High Street in either place without meeting a Mercer or one of their relatives. Nicholas was originally a lessee, but became sole owner in 1816. By then the mill produced millboard, which was its principal product throughout its working life. By 1876 it was reputed to be the largest mill for this commodity. In that year reference was made in Thorne's *Handbook To The Environs Of London* to the 'large heaps of ship rope soaking up clay' observable in its vicinity, along with 'stacks of old book-covers imbibing air and moisture preparatory to reissues'. The Mercer family remained in charge of the mill, Daniel Mercer operating it in 1839 and Daniel and Richard Mercer from 1845 to 1886, until 1893 when they disposed of their interest to the West Drayton Millboard Company. Under new management it produced 20 tons of cardboard for bookbinding each week and employed 40 local men. In the late 19th and early 20th centuries a series of disastrous fires occurred, one of which, in 1913, completely gutted it. This did not mean the end of the business, which carried on for a few more years.

In 1905 the brick mill with its tiled roof was still almost entirely water-powered, with three wheels driving four beating engines and other machinery, through there was also a steam-powered board-making machine. The business closed c1923 after which the buildings stood derelict until just after the Second World War, when most of them were demolished, having become dangerous. In 1948 the remainder were acquired, fittingly in view of the mill's earlier use, by the locally based Penguin Books Ltd. They used only the more modern part of the buildings, as a factory and warehouse; the surviving 18th-century portion remained in ruins and has progressively disappeared. In 1976 Penguin sold the site to a developer, who proposed a scheme involving the restoration of one of the wheels, part of which still remained, to working order, and this received outline planning permission. Attempts by Hillingdon Borough Council the following year to demolish the surviving northern and southern walls of the mill, in order to construct a new bridge over the river, led to a public enquiry in which the West Drayton Local History Society, opposing the demolition, stressed the prominent part the mill had played in the history of the area and the employment it had given to many local people. The Society described the proposed demolition as an act of vandalism. Reference was made to the proposal for restoring the waterwheel (which in the end was not carried out). The Secretary of State allowed the demolition of the northern wall but refused it for the southern, which was eventually pulled down in 1995.

EDMONTON

(1) Domesday Mill (see above).
(2) Salmon Brook. Early mill, later Sadler's Mill.
(3) Pymme's Brook. Weir Hall Estate.
(4) Pymme's Brook. Tanners End. 329925

In the late 12th century a watermill at Edmonton belonged to William, son of Fulbert. By 1204 it has passsed to Roger Fitz Alan who in that year granted it to John Bucointe.

Bucointe leased it to Gundred de Warenne some time before 1224, and in 1275 it was part of the fee paid by Lawrence de la Ford to Clerkenwell Priory. It may well have been the mill known as 'Scerewes mill', which stood in 1256 near Pymme's Brook. It was leased by Nicholas Roldsby to William Caton, a tanner, in 1577 along with the watermill and millhouse in Nun's Field. In 1691 a copyhold watermill, in the ownership of Lord Burghley, Elizabeth I's Chancellor, stood to the south of Bury Street. It was sold to Edward Nowell in 1613 and is not heard of afterwards, although the site continued to be known for some considerable time as Sadler's Mill (Sadler having been a late 16th century tenant).

In 1605 another watermill stood among ponds and osiers on Jasper Leake's freehold estate at Weir Hall; it was in existence in the early 19th century when it was alienated from the rest of the estate, but had disappeared by 1851. Millbrook Road, which runs from Croyland Road to Bury Street, is thought to mark the site. A further mill was that at Tanners End, known together with the Edmonton windmill, with which it operated in conjunction, as the Huxley Flour Mills. This was standing in 1791 when Benjamin Heal of Tottenham High Cross, mealman, insured the stock and utensils he kept stored there for £1,000. It was operating in 1866, but thereafter nothing more is heard.

ENFIELD

(1) Domesday Mill (see above).
(2-7) Various mills from the 13th century to the 17th. All on River Lea with one exception.
(8) River Lea, Ponders End. Replaced by (9) c1650.
(9) River Lea, Ponders End. Replaced by (10) late 18th century.
(10) River Lea. Ponders End. Standing today.
(11) River Lea. Oil mill near Ponders End. Standing late 17th century.
(12) River Lea. Paper mill. Enfield Marsh. Standing late 18th century.

Geoffrey de Mandeville, Earl of Essex, who died in 1444, granted a mill in Enfield to the Abbot of Walden, and Roger de Canteloup held one there in 1234-5. Fulling and corn mills, the former one of only two of its kind then standing in the London area, were in 1289 owned by the Abbot and tenanted by Richard de Plessis. They later became part of the manor of Durants, and were probably the ancestors of the two standing there in 1614, when Sir Robert Wroth was the lord.

One of the Wroth family's watermills stood near Enfield Lock and gave its name to Mill Marsh. In 1653 the authorities were to negotiate with John and Henry Wroth about the use of this mill for making gunpowder, but there is no evidence that anything came of it.

Ponders End Mill — 362956

The other Wroth mill stood in South Marsh, and in the late 16th century was a two-storey building with a tiled roof. In 1572 the tenant miller was Jacob Harlow. The mill then consisted of two sets of machinery under one roof. In 1650 a charter was granted by Act of Parliament for a new mill; this was known as Flanders Mill until the 18th century and from then until about 1850 as Enfield Mill, since when it has been thought of as Ponders End Mill.

The freehold of the mills at Ponders End was purchased from the Crown in September 1671. They came into the hands of James Cowper by his marriage to Ann Wroth. He leased the property with some 15 acres of meadow for 35 years to Joseph Cornish at an annual rent of £100.

Higher up the river, a Mr John Cornish was the owner of an oil mill. A few years earlier, in 1665, this had been a gunpowder mill, according to the hearth tax.

In 1671 the Ponders End mills were in the occupation of Charles Whitehead and Nicholas Ware and housed a corn mill and a leather mill; the latter, suggests local historian David Pam, probably used the oil from John Cornish's oil mill. The owner in June 1678, Spencer Cowper of London, leased the mills to Thomas Flanders for 21 years, but Flanders died that same year and the lease was assumed by his widow, Jane. Ponders End mills were described in a settlement made in 1739 in preparation for the marriage of Katherine Flanders, the then tenant, to John Clarke of Tottenham; they comprised 'two watermills, one leather mill and two tenements. The two houses . . . on the site stand one at each end of the mill'. The wording suggests the leather mill was powered by some source other than water.

In 1759 the mills, both used for grinding corn, were bequeathed by the then owner, Francis Craiesteyn, to Charles Smith. John Clarke was followed as miller by George Berners, whose lease expired in 1784. The next tenant was Augustine King George who, under the terms of his lease, had to spend £2,000 on repairs and improvements. According to an inventory they had two pairs of French stones, 4 ft 6 inches in diameter, three flour machines, one 'corn machine' and a bolter. In 1789 the structure was completely rebuilt; unfortunately, no sonner had the work been finished, just after eight o'clock on the evening of 4 March, than all the floors fell in from top to bottom. The workmen had just been paid off and only one man was in the building, his skull laid bare by the accident. 100 sacks of flour and 80 quarters of wheat fell into the river and were ruined. The collapse was attributed to the great weight of all the corn and flour inside.

In 1845 an inventory was taken of the two corn mills by John Caston, millwright of Romford in Essex. The leather mill had disappeared. The equipment of the second mill is of a far more primitive quality than the first, with greater use of wood; it was probably soon closed, since there is nother thereafter to suggest two separate mills within the same structure.

First mill — 'Ground floor: Waterwheel with iron arms, rims and shaft, iron spur pit wheel [wood cogs], iron pinion to work De [De = the same], iron bevel pit wheel [wood cogs], 4 iron pinions to work De, four iron bridge trees with lifting and lighter irons, iron upright and horizontal shafts with iron bridge trees to De, four meal troughs, and set of wheels, racks and pinions to lift the wheel shaft.

'Stone floor and upper floor. 4 pair of French burr stones 4″4′ in diameter with spindles, irons, casings, hoppers complete — iron crown wheel (wood cogs), 2 iron pinions to work De, 2 iron horizontal shafts, 2 iron rigges to drive flour machine and sack tackle, four wood riggers to drive grindstones, smut and flour-sacking machines, bolting mill, flour machines, revolving screens, and sack tackle.

'Second mill — Ground floor: Wood water wheel with iron shaft, 2 floodgates with wheels, racks and pinions to work De. 2 pairs of iron pulley-blocks with ropes to lift trapdoors, iron bevel pit wheel (wood cogs), iron wallower, iron spur wheel (wood cogs), 3 iron pinions to work De, 3 iron bridge trees with lifting and lighter irons complete, wood upright shaft, wood bridge trees, 3 meal troughs and set of wheels, racks and pinions to lift wheel shaft.

'Stone floor and upper floor.

3 pair of French burr stones 4ft 6in. in diameter with spindles, irons, casings, hoppers, etc., complete, wood crown wheel, one iron and one wood pinions to work De, 2 wooden horizontal shafts, 3 wood riggers to drive machine and bolting mills, 1 iron rigger to drive sack tackle; 2 bolting mills, flour machine, revolving screen and sack tackle.

'The whole of the machinery complete with all spindles, brasses, carriages, straps, bands, cords, pulleys etc., as fixed.'

Little is known about the mill or its owners, apart from its being rented in the 1830s by Joseph and William Farmer, until 1853 when Sir Charles Cunliffe-Smith of Saltons, Essex, who had acquired the property two years before, decided to sell it. According to the sale plan it had two pairs of undershot wheels driving seven pairs of stones, three pairs of 4′4″ and four pairs of 4′6″ in diameter. It could grind 500 sacks of flour weekly; there was storage for 1,000 quarters of wheat and barges could come in from the Lea Navigation and draw up alongside.

The premises and land were bought by the East London Waterworks for £15,750, the mill fixtures and plant valued at £1,300 18s 6d.

In 1867 a new chapter began with the arrival of George Wright. He was born at Castle Farm, near Hitchin, Hertfordshire, in 1842. After a disagreement with his father he left his farming apprenticeship to become a miller. He was first apprenticed to a watermill at Whetwell, near Luton, afterwards moving to Lensford mill near Welwyn, where he is said to have carried a sack of flour weighing 280 pounds from the mill to the local church, a distance of two and a half miles, for a wager. He came to Ponders End at the age of 25, and entered into partnership with the then miller James Young, taking up residence in one of the two mill houses. In 1870 Young died and the business was to be run by the Wright family from then on, initially in the name of Young & Wright.

In 1880 the premises were extended and partly rebuilt, with a further five pairs of millstones, driven by a steam engine. Now Mr Wright's three sons were entering the business, which was called G. R. Wright. At the turn of the century the millstones were replaced by a roller plant, except for those required to grind the company's speciality product, Imperial Wholemeal, which remained for a number of years. The mill ceased to function as a watermill in 1909 when the Lea was diverted to serve the new King George's Reservoir, and went over to electric power. In 1968 one complete set of stones and the bedstones of five more, along with the vertical shafts which drove them and the feeding apparatus for another, were still in place. Since then internal alterations have removed all trace of the original equipment.

Together with one of the 18th century mill houses, the mill building stands today in what, despite the presence on the site of several large metal and concrete silos, gives the impression — enhanced by the river with its lock and a wooded area nearby — of being a tranquil 'olde worlde' oasis, hemmed in by modernity, with a flyover on one side and two massive tower blocks on the other. Unlike the other surviving Middlesex watermills, which are wholly or largely brick-built, Ponders End is entirely weatherboarded, with the exception of the brick first floor, and white-painted. There is a lucam, flat-topped and therefore less aesthetically pleasing than some. Its one window has been boarded over, giving it a blind appearance. The mill is a large affair, with four storeys in all, including the first; the second and third are within the weatherboarded portion and the fourth the roof.

G. R. Wright & Sons have done their best to preserve the character of the mill and its surroundings, in so far as this is consistent with the efficient operation of the business. A covered loading bay has had to be constructed in front of the mill, and several large extensions built, but it is still recognisable as the structure of the turn of the century.

The firm's philosophy was expressed in a letter by Mr K. R. Wright to the *Enfield Gazette* in 1975, which he wrote to reassure local people worried about an application to remodel and extend the premises:

'We are fully in favour of the conservation of historic buildings and sites for posterity where it is feasible and sensible . . . much thought has gone into the general layout of our application so as to ensure that the buildings are situated in such a way as to be of as little detriment to the view of the buildings under preservation orders as possible.

'It should be noted that the preserved buildings are grouped on the south side of our premises while the proposed new buildings are on the north side amongst the industrial-type strucutres already there.

'As stated on the application, it is not expected that the new buildings will be over 45 feet high, which is more than can be said of the massive council tower blocks that now overshadow us on the west side of the railway!

'Flour milling has been carried out at Ponders End Mills for hundreds of years and we would like to see that tradition continued. Each generation has had to develop and modernise as time has passed and we must do the same to remain a viable business.

'We feel it is sensible for our company and the community to safeguard the livelihood of those employed here, and for us to remain a contributor to the rates and not a charge thereon. To do so we must invest in new plant and new buildings.

'We . . . feel that we are presenting a plan which would not interfere with the preserved buildings but at the same time would help to ensure that we have a working mill on this site in the future.'

Enfield Marsh Mill (paper mill)

This paper mill, which stood near Brimsdown, appears to have been short-lived, for only one reference has been found. In 1776 it was insured by John Lightly, paper maker, of Thames Street in London.

Finally, John of Enfield held a mill there in 1329 while another, thought to be the same, was granted along with the manor of Worcesters to Princess Elizabeth (later Elizabeth I) in 1550. Humphrey de Bohun, Earl of Hereford, had one in the area in 1363. An apparently short-lived mill stood by New Pond in the Chase; it was rented from the King by John Witherings in 1635 but had disappeared by 1686.

FINCHLEY

Dollis Brook. Early mill.

In 1430 a piece of land in Finchley was called Millcroft, and in 1764 two closes known as Millfields, bounded by Dollis Brook, Hendon Lane, and the Bibblesworth demesne lands suggested the presence of a watermill on the Brook.

FULHAM

River Thames. Early mill.

In the 14th century a mill stood here on the extensive lands owned by the Bishop of London. In 1307 and 1316 there are references in the manorial records to the 'mill ditch' and in 1355 to 'the pond of the mill'. In 1464-5 the mill and mill house were derelict and unoccupied, and there is no evidence that the former survived for long after this date, but this part of the river continued to be known as 'Millbank' for some considerable time.

HACKNEY

(1) River Lea. Temple Mills.
(2) Hackney Brook. Tobacco mill at Homerton.
(3) Hackney Brook. Silk mill at Homerton.
(4) Hackney Brook. Silk mill.
(5) Hackney Brook. Wick mill.
(6) River Lea. Lea Bridge Mills.
(7) Hackney Brook. Near Green Lanes.

Temple Mills. 374856

These mills, situated a little to the south of the Lea Bridge, were once owned by the Knights Templar. There were three within one building, and the mill was used for a variety of purposes including, besides the grinding of corn, pin making, calico printing, and the boring of tree trunks to make water pipes, in the days before the use of metal was perfected. One of the mills was known as Ruckholt mill (the name was still applied to the building in 1834, although by that time it had been corrupted to Rochott), as evidenced in a lease of 1601 by Edward Ryder, Citizen and haberdasher of London, to George Bromley, yeoman, of Ware. This lease also included 'a peece of ground where a lyttle leather mill sometyme stoode with the watercourse thereunto belonginge'.

One miller involved in the calico printing process was Robert Dymoke, who in 1811 was mentioned in the *London Gazette* as a prisoner in the Castle of Lincoln, though we know not for what.

In 1834 the mills were purchased by the East London Waterworks Company; they were still standing a few years later, but at some point were demolished.

(2) Homerton, tobacco mill

A watermill for the production of tobacco existed at Homerton in the mid-18th century. Part-brick and part-timber it stood beside the road leading from Hummerton (as the district was then called), to Old Ford. In 1756 it was owned by Thomas Hills and the structure and machinery together were worth £400. Two years later Samuel Hill was in possession. There were two sets of machinery, one driven by water and the other by a horse. The mill seems to have flourished briefly, for there is no evidence of its existence before or after these dates.

(3) Homerton silk mill. 365848

A silk mill is marked on the 1844 Ordnance Survey and Crutchley's map of 1856. Its site appears to be where Cadogan Terrace meets Wick Road, not far from the church.

(4) Silk mill. 362848
This mill stood on the Hackney Brook. Greenwood's map of London of 1824-6 places it at the east end of Cassland Road. It is first heard of in 1783 when owned by William Morver and Thomas Margrave. Of both brick and timber construction, like the tobacco mill and Lea Bridge mills, and with a tiled roof, it was worth £2,000 when insured against fire that year. Lyson's *Environs Of London*, published in 1811, at which time the mills were owned by Levy Smith, states that the business was 'conducted on a larger scale than any in this country'. There were two basic stages to the process: 'throwing', by which the silk was prepared from the raw state and fitted out for the loom, and 'craping', in which it was dressed and finished for the weaver. Six or seven hundred weavers of both sexes were employed, and the 30,000 spindles were worked by two steam engines 'of improved construction'. By the time of the Parliamentary Return of Mills and Factories in 1838, business appears to have declined drastically and the mill probably ceased work soon afterwards.

(5) Wick Mill
In 1766 the owner of this mill was William Walters. He went bankrupt in 1774, and five years later the mill was the property of Jonathan Loveday (1779). Until then it was a paper mill; nothing suggests it functioned as such after this date, and in an 1845 directory Henry Beckwith is listed as corn miller under Hackney Wick. Some time afterwards the mill ceased operation. In 1842 this was one of two mills at Hackney, the other being Temple Mills; both formed parish boundary marks.

(6) Lea Bridge Mills. 356867
This mill is first heard of in 1772 when Benjamin Ardley of Bow insured his utensils and stock inside it; it was then of timber and tiled construction. In 1776 it was in the possession of Jonathan Rogers and Charles Hammerton, who insured it for £2,000, along with the dwelling rooms and offices adjoining and a kiln. The notice in the *London Gazette* describes it as brick and timber built. By 1782, when a new policy was taken out, the value of the mill had risen to £4,000.

Later that year the mill passed to Samuel Lewin and Robert Thomas, mealman. Edward Phillips and William Foster had succeeded by 1786. The next miller, Richard Rogers, went bankrupt. He was followed by Charles Homerton, there in 1792. In 1791 the mills were rebuilt after a fire.

By then a second kiln had been added. In 1809 John Shepherd Killick, who had taken over from Homerton, went bankrupt, and the following year his interest was sold, passing to Thomas and Robert Tweed. They took out their own policy, one of the clauses of which prohibited the installation of a steam engine while another, interestingly, forbade there to be more than seven pairs of stones. The value of the mill had risen steadily over the years and was now £2,000.

In 1829 the mills were purchased by the East London Waterworks Company, who in 1833 dismantled them and four years later, in 1837, erected a new mill with two large waterwheels and four pumps for raising water, capable altogether of some 50 horsepower. In *Antiquities of Hackney* (1832) it is stated that prior to the mills' purchase by the company they had already been partly used for pumping.

(7) Hackney Brook

A watermill on Hackney Brook near Green Lanes is indicated by the name Millfield, by which the land in Hornsey south of the brook was called in 1577.

HAMMERSMITH

River Thames. 13th century mill. 224784

In 1448 a watermill stood here near a wharf. This may have been the oil mill from which Oil Mill Lane, which runs south from St Peter's Road to the river, derives its name.

HAMPTON

On the Longford River to the west of Uxbridge Road. 131719

In 1808 Allen Anscombe was granted permission to build a watermill 'to perform grist work for the inhabitans of Hampton parish on two days of every week'. Why it should function on this limited basis is not apparent, unless the work was being split thus with the village windmill (with which it was later run in conjunction). The first known miller was John Baker in 1823-4; John Young was there in 1845. By 1862 the mill was owned by the Deeks family who for a short time ran both it and the windmill. The miller in 1863 was James Deeks, also the Parish Overseer.

It is shown as a flour mill on the 1863 Ordnance Survey map, but latterly its purpose was to pump water to Hampton Court for domestic use. At some point it ceased to provide the main supply, continuing to serve the cloakroms for a short period up to the last war. During the 19th century, probably as the same time as its conversion to water pumping, the mill was largely rebuilt, losing all trace of its older origins. Whether anything remains today is not clear: the site is screened by a large fence.

HANWELL

River Bent.
(1) Domesday Mill (see above).
(2) 14th century mill.

In 1339 there was a mill on the Abbot of Westminster's lands here.

HANWORTH

Mersditch (later Longford River?)
(1) 14th century mill.
(2) Near Hanworth Bridge. Standing 1650.

In the 14th century there were two watermills at Hanworth driven from a dyke called Mersditch, which is possibly indentifiable with the later Longford River. The first mill was standing in 1303; its location has not been established. The second was thought by Kenneth Reid to have stood near Hanworth Bridge. It was still standing in 1650.

HAREFIELD

(1-2) Domesday Mills (see above).
(3) River Colne. Black Jack's Mill. 044904
(4) River Colne. Copper Mills. 041913
(5) River Colne. Corn mill, standing 18th century.

Black Jack's Mill

The most westerly building in Middlesex, this is sometimes assumed to be one of the three mentioned under Harefield in Domesday, but the evidence suggests it owed its construction to the Grand Junction Canal, which would have brought it a substantial trade. No mill is shown on a surveyor's sketch of the site in 1699, or on a map drawn up for the Canal Company in the 1790s. It was built in or shortly before 1797, when the canal authorities heard of its construction, served by a cut from the Canal dug without their permission. The owner was George Cooke of Harefield Park. The name is a subject of controversy; one theory is that Black Jack was a negro servant of the Derby family, who gave him his own mill. Another suggests he was a renowned bad character, with a donkey, tethered on an island in the mill-stream, which he treated cruelly. A weathervane on a nearby building shows him hauling the reluctant animal, laden with sacks of corn, towards the mill.

In 1873, the mill was offered for sale along with the nearby copper mills, as a 'Capital water corn mill called Jack's Mills, with . . . land, fisheries, cottage etc'. It was then a paper mill. The following year it was on the market again. It was large, with five floors. There were three pairs of stones driven by an iron waterwheel 15 ft in diameter and 10ft wide. It was to be let at £70 pa.

It continued as a corn mill until c1900; then it ground sawdust for the manufacture of linoleum. At some time it was used to make vegetable mats; it ceased work around 1908 and at roughly the same time was extensively damaged by fire; several conflicting dates have been given for fire and closure, and possibly one was the cause of the other. The remaining parts were incorporated in a private house some time before World War I, and a restaurant was later added. In the 1970s the wheel mechanism remained but the wheel itself had gone, although a small replica had been installed. A few millstones survived as garden ornaments.

Copper Mills

In 1674 a paper mill at Harefield was described as 'lately built'. It was mentioned again nine years later when it was rented from the Newdigate family. It was one of two mils which seem to have existed until at least 1710. Rocque's map of 1754 shows it to be where the other, a corn mill, formerly stood, the latter having moved further downstream. In 1752 the mills were sold by the Newdigates and became part of the estate of Sir George Cooke.

In 1781 the paper and corn mills were let to the Mines Royal Company who converted them both to copper mills in 1803. In their new capacity they benefited enormously from the coming of the Grand Union Canal, which made transport of the copper from Glamorgan, where it was smelted, to the Harefield mills where it was rolled, much easier. The mills were much enlarged under the direction of Robert George Spedding. Most of the copper was used to sheath the bottoms of boats although it is reported that the copper ball at the top of St Paul's Cathedral came from Harefield. The mills acquired quite a reputation, John Hassall describing them as 'the celebrated copper mills' in his *Tour of the Grand Junction Canal.* They, along with the United Asbestos Works later built nearby, were the principal employers in Harefield in the 19th century. Despite the Factory Acts, labour conditions were harsh, the number of hours worked limited only by human endurance. Production continued even on Sundays, and holidays were unknown, with the exception of Christmas Day. In 1803 about one-eighth of the village population was employed there.

Gradually, iron boats, which did not need copper sheathing, replaced wooden ones and business declined, until in 1863 the Mines Royal Company, which owned the Welsh copper mines, ceased working, the Harefield mills following.

They remained disused until Mr Thomas Newell, a Baptist clergyman who had previously run a paper-making concern in Paris, bought them in September 1870 to produce superfine quality paper. He introduced French methods and French workmen. Unfortunately this was not a success, largely because of poor discipline among the staff, despite Mr Newell's keen interest in their welfare, among other things building a lecture hall for their benefit. The mill was put up for sale in July 1873 but, as with Jack's Mill, on the market at the same time, no-one seems to have bought it. Production continued until Mr Newell faced bankruptcy, with nett debts of £7,000. This forced another attempt to sell the mills. A notice of auction in February 1879 read 'Harefield paper mills — waterpower, machinery etc., capable of producing 12 tons of writing per week also including store, office, engine house, envelope factory, lecture rooms, two steam engines, two Cornish boilers, 24 acres, two roods, 10 perches — unexpired portion of lease 44 years'. Three years later the mills were leased to United Asbestos Company, an amalgamation of two Italian, one Scottish and one English firm, with the object of bringing asbestos into commercial use. Motive power came from nine waterwheels, and the mill also produced millboard and textiles. The company secured its first contract in 1884. Nine years later amalgamation took place with Bells Asbestos Company. After the First World War, Bells sold the factory to Turner and Newall who shortly afterwards closed the mills and moved to Kent, much of the machinery going to their other works at Trafford Park, Manchester. This caused much unemployment; by the time the works had completely closed in 1930-31 Harefield village green was said to be full of men doing nothing.

Today the mill building still remains within the industrial estate at Copper Mill Lock.

HARMONDSWORTH

River Colne.
(1-3) Domesday Mills (see above).
(4) Standing from 12th century to 17th. Two mills within same building.
(5-7) Paper mills. Operated until 18th century.
(8) Calico mills. Gone by mid-19th century.

In 1293-4 there were only two watermills in the manor, one of the Domesday mills having disappeared. These both stood within the same building. In 1325 one was a corn mill and the other a malt mill. They were repaired in 1398, two carpenters working for eight days at a cost of 8s — the rate being 6d per man a day. The base of the mill, its top, the 'cogwheel', the waterwheel, a bridge and a beam needed attention. Other expenditure included 14d on four planks and 5s 10d for a 'ryne', a 'spyndele' and eight stirrups, partly of old and partly of new iron. A hopper was constructed from a cask worth 2s 3d on which two coopers worked for two days, earning 20d. They used four hoops costing 8d and nails for 2d. At this time the mill was let at £6 13s pa. By 1406 the rent was £8, and included a fishery. From time to time eels became trapped in the mill weirs, and could be used to pay the rent as well as those supplied by the fishery. This also seems to have happened at Hillingdon, West Drayton and Stanwell.

Both corn and malt mills were standing in 1433-4. Two mills within the same building were sometimes referred to in the singular and sometimes in the plural. Only one is mentioned in 1451 (when it was rented at £10), but in 1501 Harmondsworth mills, which stood at Longford and comprised a wheat mill and malt mill then in separate buildings, and the fishery were leased by Winchester College to John Smallbroke for 20 years at a rate of £10. Smallbroke agreed to re-thatch the roof in straw and to repair the walls up to a height of seven feet. He also agreed to repair the banks of the mill ponds. In 1543, when the manor was exchanged with the King, there appears to have been only one mill, that mentioned in the grant to William Paget in 1547, although in the preceding year the King had leased Longford mills for 60 years to Sir Philip Hoby. Hoby sold the mill in 1548 to Sir Thomas Paston; in 1560 Edward Fitzgarrett, who had married Paston's widow, sold the lease to Thomas Warde. It then passed, in 1565, to Edmund Downing and in 1566 to John Tamworth, who in turn sold it to Arnold Lumley in 1568. Longford 'mills' are again mentioned in 1583 and 1587. A lease of 1615 to Thomas, Lord Knyvett and William Steere of Loughborough, Leicestershire, uses the plural, and in 1622 a lease of Barnards manor farm granted to Thomas Northcott included two corn watermills, together worth £25, and one formerly a fulling mill, in one building. By 1647 the house described as having been a fulling mill had been converted into a malt mill. The wheat mill, 'furnished with French stones', stood on Colney stream, and was called Longford Mills. Later references at Longford probably refer to the paper mills.

In 1636 Longford Mill was among the Middlesex paper mills closed temporarily because of the plague. It must have later been rebuilt, for in 1647 there were three newly-built paper mills adjoining the wheat mill, one driving 24 hammers and the others 15 hammers each. In 1656 Thomas Holland agreed to repairs and enlargement. In 1662 they were let at £33 10s, and in 1694 valued at £90. In 1697 the 'mills, Houses, outhouses, wharfes, waters and premises are very ruinous and not only stand in need of being repaired, but doe also require serverall additions and alterations to make them fitt for use in the manner and forme as in and by Schedule thereof to these premises annexed is particullarly mencioned and described'. The necessary work was carried out that year by Edward White of London for £250, and new machinery to make 'fine white writing paper' installed. In the same year the two writing-paper mills were let to Nicholas Faulcon of St Giles', Cripplegate, at £65 pa, and the lease renewed in 1701. The mills stood on opposite sides of the river, one 'next to Longford town'. Further repairs occurred in 1704, and between January 1705 and May 1707 the mills sold slightly over 8,576 reams of paper. After 1762 there are no refernces to any mills until the mid-19th century, when a map of the Colne and its tributaries showed the site of former calico mills at Longford.

HARROW

River Brent. Early mill.

This mill stood on the Brent at Alperton, and gave its name to a field called Mill Acre. It was in existence by 1236. Milling on the Brent was a hazardous business due to frequent flooding; in 1273-4 repairs to the watermill included raising timber 'submerged by the water'. This seems to have led to its complete replacement by the windmill; an account for repairs carried out to buildings on the manor in 1348 mentions only the latter, which may indicate that the watermill was no longer in use. It had definitely gone by 1470-71.

HILLINGDON

Frays River. Cowley Hall Mill. 049819

In 1769 this mill was occupied by Thomas Dagnall Senior. Insurance policy notices described it as 'timber and tiled', the tiled part most likely the roof. In 1783 and 1804 the tenant miller was Richard Biggs. By 1783 part of the structure had been rebuilt in brick, in common with many watermills of the period. Mark Burr was the miller in 1823-4 and 1832-4. He was succeeded by William Burr, who combined the duties of miller with those of a commission agent and was declared bankrupt in 1847. In 1851 and 1862 John Benbow was in charge; in 1856 he was assisted by a relative, Clifton Benbow. The mill was destroyed by fire in February 1864.

HORNSEY

14th century mill.

A 14th century lawsuit mentions a 'Stowelmyll' attached to the Manor of Farnecroft, later Farnefields. It stood close to the Islington mill and the two may be the same, unless the 'Stowelmyll' stood on the Moselle stream of which 'Stowell' may be a mispronounciation.

HOUNSLOW

(1) River Colne. Flax mills, standing 19th century.
(2) River Colne. Paper mill, standing 1680.

Flax mills

These mills, on the River Colne and used to manufacture flax into yarn, first appear on the 1816 Ordnance Survey map; their last appearance is on Davies' map of 1859, which shows them close to the powder mills. Along with many other local mills they were owned by the Duke of Northumberland. The lease was sold in October 1817 after the owner went bankrupt. The property included warehouses, hackling shops, a dwelling house, several cottages, drying houses and stables. The lease was again sold in 1820, after John Petrie and John Ward, who ran the business under the name of the Flax Mill Company, had been declared bankrupt. The mills were worked by two waterwheels of 16 feet diameter.

Paper mills

A visitation of the plague in 1636-7 led to the Priory Council on 12 September directing the JPs of Middlesex to 'search for and burn the contents of all rag shops and to hinder the gathering of rags near London so as to prevent their being transported or used for paper, in order to prevent the spread of the contagion'. These instructions may not have been effective, for six days later an order was issued to close all paper mills in the county. Later, in 1636 William Bushee had to answer at the Middlesex Sessions for disobeying the order and 'grinding ragges in his paper mill that came from London, whereby one of his servants became infected with the plague'. Bushee's mill is thought to have been that shown on Seller's map of 1680. It is not clear how long paper-making continued at Hounslow thereafter.

ISLEWORTH

(1-2) Domesday Mills (see above).
(3-8) Various mills from 13th to 17th century.
(9) On site of Kidd's Mill. Erected 1543. 165759
(10) On site of Kidd's Mill. Rebuilt 1669. 165759
(11) On site of Kidd's Mill. Burnt 1795. 165759
(12) Duke of Northumberland's River. Kidd's Mill. 165759
(13) River Thames. Good Intent Mill. 167754 (approx.)
(14) Duke of Northumberland's River. Brazil Mill. 158760
(15) Duke of Northumlerland's River. Calico mills.

In the 13th century a mill was built on the manor of Isleworth by Richard of Cornwall, brother of Henry III, to whom the manor had been granted. In the 1264 rebellion against the King, Richard supported his brother and thus aroused the wrath of the citizens of London, who came to Isleworth to destroy his possessions there, the mill included. The mill was presumably rebuilt, for one is recorded in the manor in 1297.

The manor was thenceforth granted to each Prince of Wales, as Duke of Cornwall, for life, until this was revoked by statute and the lands given to Syon Monastery. In 1352 the manorial mill stood near the manor house on a stream called the Bourne, which followed the course of the present Duke of Northumberland's River. In 1362 John Weeke was steward of the Manor, and upon his report, the bailiff was commanded to rebuild the watermill. In 1370, according to a survey by Adam de Herlyngdon for the Treasury, there were two watermills attached to the manor; one needed rebuilding, at a cost of 40 pounds.

The Bourne seems to have had defects as a mill-stream in the 14th century and by 1463 the Isleworth mill was disused and derelict; shortly afterwards the millstone was sold. By 1506 the derelict millhouse had been leased with the adjoining manor house to John Fox, Bishop of Winchester, who had granted Syon Abbey another mill, built on the Crane between Twickenham and Isleworth some years before, in exchange.

In the time of Elizabeth there were two watermills in Isleworth, worth £20 pa. The Crane was little more effective than the Bourne; the new mill between Isleworth and Twickenham is not referred to again.

Kidd's Mill

By 1543 another mill was being built on the old site near the mouth of the Bourne. It and its successors remained an appurtenance of Hounslow Manor until 1876. It had ten millstones in 1553, of which four ground corn and the fifth wood dyes. It was a corn mill only in 1669, when it was to be rebuilt by the lessee. This mill stood on the Duke of Northumberland's River at Mill Bridge in Church Street, near what is now the old part of Isleworth. The successive mills on the site were owned by the Duke of Northumberland and the mill was known as Manor Mill until well into the 19th century. In 1772 the mill was owned by Edward Merchant, but George Merchant was in possession in 1788. It was destroyed by fire in 1795 and rebuilt. The original structure was of oak on a brick lower storey and had a tiled roof with three dormer windows.

In 1804 Messrs Leader, Attlee & Co of Wandsworth spent the curious sum of £29,999 19s 9d repairing and carrying out modifications, in return for which the Duke leased the mill to them for 60 years. A few years later the lease passed to Benjamin and Robert Kidd, whose partnership was dissolved in 1838 after which they sold the mill to their brother, Samuel.

'Sammy' Kidd was, one contemporary newspaper wrote, a man of impeccable character, jealous of his good name as a businessman, a kind employer, and always mindful of the effect of the business on fellow residents. The mill hands had to start work early, and work night shifts, so that barges containing corn or meal could come under the road bridge into the mill basin to be loaded or unloaded whenever the tide was right (and regardless of the weather, a sackcloth canopy attached to one of the lucams giving protection against rain and wind).

It became the practice to 'knock them up' at night and early in the morning. This so disturbed inhabitants' peace and quiet that they complained to the Churchwardens.

Soon afterwards the following public notice was issued:

'To Millers and others,
In consequence of the continual complaints that are made of the annoyance caused by the Millers, &c., calling up their mates at all hours of the night, and bawling in the streets so as to disturb the inhabitants of the village, Mr Kidd has promised to dismiss from his service any person who shall repeat the offence after this warning, and those with whom they lodge are requested forthwith to have bells put up in the rooms they occupy, which will at once remedy the nuisance. The Millers are recommended not to take lodgings where they cannot get this indispensable accommodation.
(Signed) Horatio G. Day
Samuel R. Goodenough
Churchwardens
March 6th 1850.'

Mr Kidd's generosity was demonstrated during the 1840s potato famine and the economic depression which followed the Crimean War, when he donated flour to poor inhabitants.

In 1846 Kidd formed a partnership with William Podger from Somerset, and at about the same time two steam engines were installed to supplement the waterwheels. The mill was then one of the largest in England. Its weatherboarded cowlings and three lucams of exceptional height made the building a distinctive sight.

In 1863 Kidd retired from the business, although it continued to be known as Samuel Kidd & Co; Podger now ran it in partnership with a Mr Odlum until the latter died in 1875. Two years later Herbert Robert Perry became a partner, remaining until the busines was made a Joint Stock Company in 1886 as Messrs S. Kidd & Co. Mr Podger then retired from active management but continued as a buyer. Samuel Kidd & Co became a subsidiary of Associated Flour Mills Ltd.

A biography and character sketch of William Podger appeared in a local paper in March 1887:

'One of the most genial and honourable of men and the best buyer on the market is a concise and literally correct description of the well-known proprietor of the Manor Mill, Isleworth; no man connected with the trade is better liked or could be more highly respected. A prominent member of the Corn Exchange remarked the other day that he had transacted business with Mr Podger for forty years without once having had any dispute or unpleasantness.

'As all can testify who have ever done business with him his word is his bond, and when once it is given there is no need of a contract "signed, sealed and delivered". He is a large buyer of the best English wheats, and much sought after by sellers of native and imported produce. There are always three or four deep waiting to get hold of him. Mr Podger attained the good age of three score years and ten in January last, but is as hale and active, and cheery as the youngest men in the market, and it would certainly take the younger men all their time to steal a march on him. We trust his familiar presence will not be missed from the market for many years to come.'

In 1887 the mill had two waterwheels and was fitted with 29 pairs of stones and 11 sets of rollers — an indication of its huge size — 'with machinery of the most improved description', and was capable of turning out 3,000 sacks of flour per week. A third steam engine had been added. In 1894 it was converted entirely to roller grinding, making it one of the most modern and up to date mills in the London area, if not the country. However, its geographical position at Isleworth, comparatively far from the Port of London, eventually made it difficult for it to compete with flour mills there, and in 1934, on the expiry of the lease from the Duke of Northumberland, it was shut down and the business transferred to the Empire Mill in Victoria Docks. The machinery was then removed and the building allowed to stand derelict until demolished in 1941.

It is unlikely that the mill used water as its principal power source right up to the time of its closure, although a small channel from the River Crane was used to drive a stone plant for the manufacture of a speciality product called 'bold metal meal'.

Good Intent Mill

This was one of a number of windmills and watermills run by co-operative societies; as such it was to play a notorious part in the political history of Middlesex.

The Good Intent Society was formed in May 1801 for the purpose of grinding corn at a cheap rate for its members' own use. Originally it had only four or five members, but by August the number had increased to 80 or 90, who held 130 shares between them. The chief officials or 'Inspectors' of the Society entered into an agreement with Richard Friday, whereby he contracted to sell them a piece of freehold land in his ownership for £300 on which to build a mill, at the same time clearing away ballast from the Thames so that barges could be brought up to it. The building was not completed until 1803; in the meantime, a General Election had been held.

At the time, all those belonging to a co-operative society were legally entitled to vote. During the course of the election membership of the Good Intent Society, which as its outset was between 200 and 300, increased considerably; many of the new members were labourers and mechanics, people who under other circumstances would have been the least able to vote.

Membership was being offered on specially easy terms, the subscription fee substantially reduced. Many people were allowed to vote the day after joining, which was not strictly legal. Another of the candidates in the election, one Mainwaring, who had polled the least number of votes, objected. The matter was investigated by the House of Commons, and 300 votes were struck off the register, while the sheriff of the county, along with the returning officer in the election, was sent to prison. Besides other infringements, the Commons noted that, at the time of the election, the mill was incomplete, with the roof not yet on; thus the Society was not fully established, which it had to be for the votes of its members to count.

The mill was standing in 1826, when Bonner in his *Guide To Richmond* commented acidly that it had been in business before it was actually complete, grinding votes instead of flour. At some time the Good Intent Society was wound up and the premises became a cement works. The mill was still standing in 1888, when a line drawing depicts it with several other buildings nearby, from which rise two large chimney-like structures, presumably connected with the cement-manufacturing process.

The mill was a large structure, of brick for the first storey and wood above. It had a 90-foot frontage on the river. The red-tiled roof had four dormer windows on each side and there was a lucam.

Brazil Mill

Norden, in his *Speculum Britanniae* of 1592, describes a copper and brass mill at Isleworth. It stood on the Duke of Northumberland's River in St John's Road and, according to the proceedings in a dispute in 1596 between the two partners, it was built between 1581 and 1587. One of them, John Broad or Brode, sought grants from Elizabeth I for the mill, which he claimed employed processes for making copper plate never before used in England, but he did not get them. Norden describes the mill thus: 'Thistleworth or Isleworth, a place scituate upon the Thames . . . Not far from whence betwene it and Worton is a copper and brass myll, where it is wrought out of the oar, melted or forged. The oar or earth whereof it is contryved is brought out of Somersetshire from Mendipp, the most from a place called Worley Hill. The carriage is by wayne, which cannot but be very chargeable. The workmen make plates both of copper and brasse of all seyces little and great, thick and thyn, for all purposes. They also make kyttles. The furnace and forge are blown with great bellows, raysed with the force of the water, and suppressed agayne with a great poyes and weight. And the hammers wherewith they worke their plates are very great and weightie, some of them of wrought and beaten iron, some of cast iron, of 200, 300, some 400 weight, which hammers so massye . . . are lifted up by an artificiall engine, by the force of water, in that altogeather semblable to the Iron myll hammers. They have snippers, wherewith they snyppe and pare their plates, which snippers being also of a huge greatnes, farr beyond the power of man to use, are so artificially placed, and full ingenious devices thereunto added, that by the motion of the water also the the snipper open and shut, and perform that with great facility, which ells were very harde to be done'.

Glover's map of 1635 describes it as a copper mill, although it is marked as a paper mill on a map of 1607. A map of 1623 calls it a brass mill, and also states it to be 'decayed'.

Unlike many of the others in this area the mill never belonged to Isleworth Manor; the right of using the water of the Duke of Northumberland's River to drive it was leased. One of the terms of the lease was that it should not be employed for grinding corn.

The mill was in operation again by 1671 when it was used both for making paper and for grinding brazil-wood for dyes. By 1721 it was a brazil-mill only, and continued as such until well after it became a corn mill in the 19th century, the stipulations covering its use rescinded. The mill was included in the will of Isaac Scott in 1752, according to which it had recently been a gunpowder mill but was a copper mill again. The largely wooden building was destroyed by fire about 1860. Then it was grinding snuff, and as this was contrary to the terms of the insurance policy the insurance company would not pay up. This ended the life of the mill. Two years later it was bought by the owners of the adjoining brewery, who demolished the remaining buildings.

A few traces of the mill, including the millstones and remains of the wheel and mill race, were uncovered in 1936 by builders excavating new paving at the side of the stores on the site of the mill house.

The mill was worked by a family named Stanborough in its latter years: Charles in 1839, A. & E. Stanborough in 1855, and E. Stanborough in 1862.

Calico-Printing Mills

These were set up on the Duke of Northumberland's River after its separation from the River Crane. The first, which stood on the east bank near Worton Bridge, was started in 1769; the other, on the west bank a little further upstream, by 1818. Both mills had gone out of use by 1833.

A print of one of the mills reveals the waterwheel to have been longer than it was high.

ISLINGTON

Near Green Lanes. Gone by mid-18th century.

The mill which stood here near Green Lanes, and whose foundations remained until c1850 in a field behind a public house called the Pegasus, does not appear on Rocque's map of 1754; the main part probably disappeared before then. Certainly no trace is visible today. The stream which drove it was probably affected by the construction of the New River. At one time, prior to the Dissolution of the Monasteries, the mill belonged to the Hospitallers of Clerkenwell.

KINGSBURY

River Brent.
(1) Domesday Mill (see above).
(2) Possible mill, 15th century.
(3) C16 mill.
(4) C16 mill.

A place known as 'Brentmill' existed here in 1461 and 1499, but there is no proof that this was a watermill, and besides it may have stood equally in Kingsbury, Hendon or Willesden. The latter is also true of a known watermill included among appurtenances of Coffers Manor in 1556. The only definite post-Domesday watermill in Kingsbury is mentioned in 1596, when Jon Chalkhill erected a mill over the Brent to the south of Blackbird Hill. Nothing is subsequently heard of it.

LALEHAM

River Thames. 042693
(1) Early mill.
(2) 18th century mill.

A watermill in the ownership of Laleham Abbey underwent major repairs in 1276 and in 1302 was moved to a new site. The mill and manor had been owned by Westminster Abbey since before the Norman Conquest. A successor was standing on the Manor in 1608, 1631 and 1644, when it was the property of Sir Henry Spiller. The most likely site for these mills is the weir to the north-west of the village.

LIMEHOUSE

Flax mill, standing 1838.

In 1838 a flax mill here made use of a waterwheel; it was of low power (2 horse), and supplemented the two steam engines, of 14 and 30 horsepower, which were the principal source of power for the works, to be used in case one or both of them should break down. A Parliamentary Return gives the number employed as 33, all adults.

OSTERLEY

River Brent. 16th century mill.

In the 16th century a tributary of the Brent ran through Osterley Park, which was owned by Sir Thomas Gresham, and here c1571 he built a water corn mill, probably replacing one erected by the previous manorial lord, the Abbess of Syon. The mill-pond, the easternmost of several on this stream, lay between Windmill Lane and the Brent. Before his death in 1579 he established a paper mill within the same building. Norden, in his account of the Park and mansion in 1593, says the former was 'garnished with manie faire ponds which afforded not only fish . . . but also great use of milles; as paper milles, oyle mills and corne mil, all of which are now decaied (a corne mill excepted)'. The corn mill does not seem to have survived for long. An inquiry was held in 1584 to decide whether Gresham's mill had encroached onto the King's highway, and its situation was described as upon the Brent near 'Cruxellesforde'. By 1574 the mill had gone but the pond remained.

PINNER

River Pinn. Love Lane. Demolished 1828. 121900

Pinner's watermill was to be relatively short-lived, leaving behind no illustrations and featuring on none of the contemporary maps. The mill was not standing in 1770 when the site was acquired by John Foster; it is first referred to in 1787 when a fire on 18 January caused damage worth £90 to the building and £20 to the stock. Foster was thus prompted to insure the mill, which by then had probably been partially restored, and its contents, along with a kiln for drying beans (a fitting some insurance policies forbade on account of the fire risk, and possibly the cause of the blaze) with the Sun Insurance Company for £50 on 3 January 1788; on 15 February he reinsured the mill for £200, the repairs completed.

In 1795 the tenant miller was Joseph Dell, whose family operated the mill in conjunction with the local windmill. The Dells relinquished the mill in 1826. In 1828 the half-acre site was sublet by Foster to Vincent White, a bricklayer and builder who pulled down the watermill and replaced it with a row of four cottages, using the bricks from the former. The lease describes it as a building 'formerly used as a watermill', indicating that it had been disused for a number of years.

The site was behind the present nos 27-35, Love Lane.

RUISLIP

River Pinn.
(1) Early mill. 081875
(2) Possible mill, standing 1442.

There seems to have been a watermill at Ruislip in or before 1248, when Roger de Southcote was paying rent for a millpond called Sitteclak. The first reference occurs in 1294, when the Abbey of Bec, whose lands included the manor of Ruislip, was taken over by the Crown, which compiled a survey or Extent of its new possession. This mentions a watermill, valued along with the local windmill at 40s. References to it cease after the early 15th century, although a mill of unspecified type, which this may be, is mentioned in a rental of 1442. The site is believed to have been somewhere near Clack Lane. In the 1970s the remains of what appear to be a mill leet were still partly traceable, starting from a pond on the Pinn near Fore Street in Eastcote and then running north of the Manor Farm to rejoin the river west of Bury Street.

ST CLEMENT DANES

Possible mill. 311812

Milford Lane, off the Strand, could well indicate the former existence here of a stream which powered a watermill. The name may equally have come from a windmill.

SHEPPERTON

(1) Early mill.
(2) 14th century mill.
(3) 14th century mill.
(4) River Thames. Standing 1805. 077675

A watermill stood on Halliford Manor in 1289. There were two newly-built watermills on the manor in 1320, so the previous one may have gone out of use. It may also be the one referred to in 1336 and 1343 as old and of little value, although this seems to have stood on the separate manor of Shepperton. Another mill, perhaps the successor to one of these, was standing in 1805; it survived until well into the 19th century and may have given its name to Millbrook House. It stood to the south of the church.

SOUTHALL

(1) Domesday Mill (see above and below).
(2) Overshot mill. Dormans Well. 143807

The mill mentioned under Hayes in Domesday is thought to have been in Norwood. The next reference to a watermill occurs in 1578 when one was leased by Robert Chamberlain, the lord of the manor of Norwood. In 1596 the executors of Anne, Lady Dacre, reserved for themselves, along with the windmill then in the manor, a watermill and garden. There was still a mill in Norwood in 1611 and in 1673 an overshot watermill was there. Mr Hamton owned or leased the mills in 1676. In 1716 the overshot mill passed to Sir George Cooke, future lord of the manor, and was offered for sale with the other Cooke properties in 1770. It still comprised part of the manorial estate in 1800, when it stood along with a house and other property at Dorman's Well. In 1819 a former miller, George Hudd, was registered as bankrupt. From at least 1862 until 1886 it was worked by Alfred Robinson, apparently assisted on and off by a relative called Herbert. Shortly afterwards it went out of use. The mill was supplied by its own pond.

Rocque's map of 1754 shows a mill standing on a small stream running through Osterley Park on the southern boundary of Norwood Parish, between Windmill Lane and the stream's confluence with the Brent. There is no other record of a mill here and, as the overshot mill is not marked, it is possible this was it, placed in the wrong position.

STAINES

(1-6) Domesday Mills (see above).
(7-9) At least three mills, one of which was in Yeoveney: the sites of the others cannot be identified.
(10) River Colne. Hale Mill. 033717
(11) River Colne. New Mill. 033718 (approx.)
(12) Wraysbury River. Pound or Mustard mill. 033718

Staines' position on the river Thames, of which it was a major crossing-point, enabled watermills to flourish there. There were six at the time of Domesday Book, if those in parts of the manor other than Staines itself are included. Later in the Middle Ages there were two in Staines town, each with several sets of machinery, and one at Yeoveney. There are a number of references to mills which cannot be traced, such as the 'Hurst Mill' mentioned in 1354. Various other mills are mentioned in conveyances of the properties in the 13th and 14th centuries, and it is not clear whether these represent a third mill or one of the above. In the 15th century one of the mills was fitted with a small plant for grinding malt. One seems to have disappeared by the beginning of the 16th century.

The following individual mills can be differentiated:

Hale Mill

During its early life the ownership of Hale Mill appears to have been divided; in the reign of Edward I (1271-1307) part belonged to John de la Hale, from whom its name is derived, while another part was in separate ownership from 1271 to 1353. After this date the mill belonged to Westminster Abbey. It was derelict and out of use from c1472 to 1490. It was probably one of two mills in existence in 1503 and 1694. Since it stood on the River Colne it was probably that whose occupier claimed to have been injured by Henry VIII's construction of the Duke of Northumberland's River.

As with many other sites it is highly unlikely that the mill standing in the late 13th century, and that recorded in 1694 and subsequently, are the same, although the same site may have been used. By 1755 the mill had passed to John Finch, also owner of Pond Mill, whose family retained it until the mid-19th century, although latterly they did not work it themselves. In 1781 John Finch insured the brick-and-tiled mill for £700. By the following year the miller was Thomas Finch, presumably a relative, who then took out a new insurance policy to include the adjacent house and offices, the overall cost of the property rising to £1,080.

By 1855 Hale Mill had ceased to produce flour and was used as a papier-maché factory. In 1864 it was taken over by the Linoleum Manufacturing Company, who bought it outright in 1871. Until fairly recently the mill buildings were still evident, but now nothing remains.

New Mill

New Mill, first mentioned in 1388, stood just to the east of the now defunct Great Western Railway station, on or near the site of the later Pound Mill. At first the mill belonged to Westminster Abbey but it was later divided among various owners and lessees. In the 15th century part was in use as a fulling mill, while in 1472-3 part had been disused and occupied for eight years, the remainder of the building presumably still in use, but in 1490 the whole was occupied. Here is another example of the practice, common in western Middlesex at the time, for two mills within the same building to be in separate ownership and used for different purposes.

Pound Mill (or mustard mill)
Pound Mill, so named after the neighbouring parish pound, is first heard of in 1747 when it came into the possession of John Finch, who eight years later was to add the Hale Mill to his properties. It was then said to be newly-erected, although there was a mill on the site in 1682 – and in 1916 part of the machinery was dated 1712. It seems likely that Finch's mill had been substantially rebuilt, incorporating material from an earlier one. In 1782 the mill, brick-and-tiled like Hale Mill, and its contents were worth £500. In 1793 it was worked jointly by Thomas and John Finch and was worth £1,000. Unlike Hale Mill, Pound Mill continued to be worked by the Finch family until it passed out of their hands. During the 19th century the mill's chief business, latterly carried on under the name of Finch, Rickman and Co, was the production of mustard, although part was still used for grinding corn. In the middle years of the century the business was a prospersous one, employing a considerable number. The owner was Charles Finch, whose grand house nearby later became the GWR station. Later, trade declined and the business was sold in 1900. It continued as a mill until 1912 and in 1916 was purchased by the LMC, who shortly afterwards demolished the upper part.

The lower part, to first floor level, was still standing in December 1945 when examined by Mr Denis Sanders of Feltham, but has now gone. The building was of red brick and 33ft wide by 24ft from east to west. A 14 ft diameter breast shot wheel was still in place on the east of the building. There was evidence that at some time a turbine had been installed.

STANWELL

(1-4) Domesday Mills (see below). Early mills.
(5-10) At least five mills known to have existed in mediaeval and early modern periods.
(11) River Colne. Poyle Mill. 033766
(12) River Colne. Upper Mill. 041751
(13) River Colne. Snuff Mill. 034741
(14) River Colne. Lower Mill, Leylands Lane. 039749

The sites for four watermills recorded at Stanwell in Domesday are unknown. A new mill appears to have been built some time in the 14th century and there were at least three in 1606. Like others in the area, the mills at Poyle and Stanwell were adversely affected by the Duke of Northumberland's River. A new mill was built later in the 17th century – possibly on the site of those standing in 1606, although four mills are recorded in the late 18th century.

In the Middle Ages several mills stood in Stanwell, none of them in manorial ownership. One of them was held by Richard Peacock, and later his son John, in the 14th century, and was still standing in the early 17th, by which time it had become known as 'Peacock's Mill'. A second mill was acquired in the 15th century by Richard Bulstrode, and one, which in view of the name may have been the same, was part of an estate belonging to Edward Bulstrode in 1598.

Poyle Mill
Of all the Standwell mills this has the longest continuous history. A mill stood on Poyle Manor in the 13th and 15th centuries; it later passed with the Manor to Andrew Windsor, to be alienated by the Crown in 1612 as part of his former properties. In the 15th and 16th centuries it consisted of two mills under one roof.

By 1636 it ws used to manufacture paper and 12 men were employed. In the 18th century it was tenanted by the Bullocks, who later acquired the freehold. The first of the family to work it, Henry, and his sons used it to dress leather as well as paper-making. The Bullocks sold the mill in the early 19th century and in 1807 it was run by William East and Richard Ibbotson. Ibbotson and his sons later became owners. The firm used it at first principally for making paper, and later leather and asbestos board, until the partnership between the Ibbotson brothers, Richard, Thomas and Percy, then carrying on the business of paper and millboard manufacturers as P. & R. Ibbotson, was dissolved in 1850. After 1890 it was used by various owners for various purposes, the latter including asbestos, fibre, flock, artificial manure and bricks. By 1956 the building had been pulled down and its site was occupied by the Wilkinsons Sword Co and another firm named Graviner, although the weir still remained.

Upper Mill

In 1385 John Donet was presented for building a mill over a common footpath in Stanwellmoor, and was ordered to remove it; but for some reason this illegal mill remained, becoming the ancestors of the building which stands today in Horton Road. In 1472 it passed to Thomas Windsor. It is almost certainly the same as that known in the 15th century as North Mill and from about 1630 as the Upper Mill. The North Mill was the property of the Crown in the 16th century and until 1612, when it was granted to Lord Knyvett. From c1610 to 1652 it functioned both as a corn and paper mill; individuals employed in paper-making included George Hagar (from 1682 to 1691) and, during the 18th century, Eustace Burnley. In 1768 John Stevens was miller and the mill and house, which communicated with each other (as they do today), were worth £150. The mill was then partly of timber and partly of brick. From 1771 it is referred to as the 'New Mill', probably due to a recent rebuilding.

The mill ceased to produce paper between 1772 and 1793, afterwards functioning entirely as a corn mill until the early 20th century, at about which time it was sold by the lord of the manor. In 1781 it was owned by a Mr Dell and in 1787 Simeon Warner Hagen. The latter was still owner in 1803 but in 1806 it was in the possession of John and Samuel Saunders. In 1813 the mill was advertised in the *Reading Mercury* to be sold by auction, the notice stating that it had four pairs of stones. Apart from the mill itself and the mill house, the premises included a stable. In 1832-4 the miller was Henry Carpenter; by 1845 he had been succeeded by James Carpenter Ward, 'J. C. & H. Ward', are given as millers in an 1862 directory, the 'H.' probably being Henry Carpenter Ward, who worked the mill in 1866 and 1890. William Wooster operated it from 1895 to at least 1917. By 1931 the mill had ceased to produce flour and was instead used in the manufacture of dyes.

From early this century water power was supplemented by steam, though it was not finally discontinued until c1950, after which electricity provided the motive power. The wheel and machinery remained in 1956 but were removed shortly afterwards. The former and parts of the latter were at some point examined by Denis Sanders. The all-iron breast shot wheel measured 16 ft by 8 ft wide and had three rims, with eight arms each of eight segments each, and 48 buckets. It was mounted on a square, iron wheelshaft with wooden struts. The pitwheel was also of iron, 12 ft in diameter with eight arms and 112 wooden teeth.

The building we see today dates essentially from c1900, when the mill was rebuilt after being severly damaged by fire, although it has been altered several times since. It is a fair-sized building of red brick, nicely built but unremarkable, its wooden lucam the only distinctive feature. Today the mill houses a chemical company, while the mill house, with which it forms an unbroken front, is still occupied.

Snuff mill

By 1781 Edmund Hill owned a gunpowder mill here; around 1820 it was taken over from his successor by Curtis & Harvey, who also operated the mills at Bedfont and Hounslow. By 1896 it has been converted to a snuff mill, but later became a corn mill, remaining as such until burnt down in 1925. In 1956 a house nearby marked the site.

Lower Mill

It is not clear whether the Lower Mill was one of the two held by the lords of the manor in the 17th century, or was built much later. The first clear evidence of its appearance is on the 1816 Ordnance Survey, represented by the words 'Corn Mill'.

Thomas Wheeler was the miller in 1832 but, from the mid-19th century until its demolition between c1886 and 1896, it was managed by the Carpenter Ward family, who operated the Upper Mill a few hundred yards downstream.

STEPNEY

Domesday Mills (see above).

STOKE NEWINGTON

Hackney Brook. 16th century mill.

In 1577 the land in Hornsey, south of the Hackney Brook, near Green Lane, was known as Millfield, which may suggest a watermill there.

SUNBURY-ON-THAMES

(1) River Thames. 17th century mills 101677
(2) Feltham Hill Stream. Kempton Park. 121702 (approx.)
(3) Fordbridge Road. Standing 1826.

A succession of watermills existed at Sunbury between the 17th century and the 20th. In 1623, 100 poor bargemen complained that Sir Thomas Lake had 'caused watermilles to be built' there 'to the great hindrance of a great many poore men which gets their living by the . . . River'. Their statement ended 'But what cares he or such as he?'. The diversion of water to drive the wheels meant that there was insufficient in the locks to enable barges to go up or down the Thames. The Lord Lieutenant of Middlesex wrote to the Lord Mayor of London in his capacity as Conservator of the Thames requesting that he suppress the mills.

The 1722 map of Sunbury Manor shows a field called Mill Acre between the present Fordbridge Road and The Creek. Its western boundary follows roughly the line of the River Ash, making it likely that the mill which stood there was a watermill.

A watermill was standing in the manor of Kempton, on the Feltham Hill stream, by the 16th century. It was rebuilt in or after 1650 by the lord of the manor. In a lease of Kempton Park by its owner Edmund Hill to William Rolfe in 1803, Hill reserved the right to pen up the water in the fish ponds there in order to supply a watermill and to lay pipes and dig a canal to convey water from the ponds to the mill. This was to be erected between the ponds and the Thames on premises then in the occupation of John Shore.

Early in the 19th century, John Gerard Colbert, a watchmaker of St Marylebone of Swiss or German origin, took out a patent for improvements in the manufacture of screws. This was later acquired by John Sutton Nettlefold, an ironmonger of Holborn, In 1826 Nettlefold issued his first price list, in which he stated that he had established a screw factory, powered by water, at Sunbury-on-Thames.

The enterprise was a great success and in 1834, to cope with increasing demand, Nettlefold took his son, Joseph Henry, into partnership, and transferred the works to Birmingham, where he formed a partnership with Joseph Chamberlain, later to be a famous Radical politician and Colonial Secretary during the Boer War.

The mill stood in Fordbridge Road, where it was driven either by the Ash or the Thames, but its exact site is not recorded.

TOTTENHAM

River Lea.
(1) 13th century mill.
(2) 14th century mill.
(3) Standing 16th and 17th centuries. Gone by 1680.
(4) Oil mill, standing early 17th century.
(5) Leather mill, standing 17th century.
(6) Paper mill, later corn mill. Rebuilt 1788, burnt 1852. 349895
(7) Oil mill. Built 18th century. Rebuilt 1788, burnt 1852. 349895
(8) Lace and cotton mill. Standing 1829.

In 1254 a watermill at Tottenham was divided, along with the rest of the manor, among several owners. One is again mentioned in 1321 when William atte Hagh received 'the property of the Lord's watermill with bakery, pond . . . 10 quarters of corn and half a mark for the fishery, the said William to maintain the structure'. In 1374 Sir Thomas Heath's share of the mill was in a ruinous condition. At a manor court held about 1523 it was reported that 'the watermill of the manor is in great decay for want of repairs by timber in the rooms and in other houses of the same mill and in grounde works, and that the ditch of the same mill is not cleared water and "le ossyers" trodden down and destroyed which formerly were of the yearly value of 20 shillings a year; and that one John Kyrton is held to repair the aforesaid mill and all the premises with appurtenances by reason of a certain demise made to one John Davy, to whom John Risley, knight, then Lord of the Manor granted the said mill . . .'.

The miller was fined for excessive tolls in 1530, and a tenant was fined for refusing to take his corn there in 1558. It was leased along with 12 acres of the nearby meadow in 1585.

In 1619 there was a leather mill here in addition to the corn mill. The site was on Mill Road, and was approached by a lane south of the later Ferry Lane. By 1622 they had been joined by an oil mill, which was short-lived; in June that year Middlesex JPs were instructed by the Privy Council to tell its owners that the noise and smells of the oil-making process offended the King; since he was about to visit the area, it would be advisable to cease their business.

In November 1656 the lord of the manor was taken to court for converting the corn mill to a gunpowder mill, to the injury of locals' welfare.

The corn and leather mills had gone by 1680, when a paper mill alone is shown on John Seller's map. It was owned in 1735 by Israel Johannot, one of a well-known firm of French paper-makers. His widow later married Thomas Cooke, an excise officer who had been apprenticed to the paper trade, and the latter became tenant. Cooke appears to have been rather a nasty piece of work, although his audacity is almost admirable. He is presumably the same Thomas Cooke who was rewarded by the Royal Society of Arts in 1763 and 1764 for his skill in making paper with copper plates. In 1770, on the expiry of Cooke's lease of the mill, a new one was granted to Edward Wyburd with permission to convert it to corn grinding. Shortly afterwards Cooke applied to remove this lease. Wyburd objected strongly, but it was only after physical violence and court action that he was able to gain possession.

Cooke had risen by his own efforts to the position of exciseman from unpropitious beginnings. He was born near Windsor in 1726, the son of a wandering fiddler who earned a living by playing in alehouses. When still an infant he was sent to live with his ancient grandmother at Norwich. As a boy he worked in a local factory; the other boys clubbed together to provide refreshments which they ate together, but Thomas had no part in this arrangement, preferring instead to retire to the edge of a brook and dine off a halfpenny loaf, an apple and a draught of water from the brook. With the money he saved he paid a boy who was a monitor at the village school to instruct him in reading, writing and arithmetic. When he reached maturity he was employed at a paper factory, pleasing his employer by his sobriety and hard work.

It became his ambition to be a customs officer, for he had been listening carefully to their conversation, and knew they were highly paid. His employer recommended him to the authorities and Cooke set off for London. Here he obtained his Excise Appointment and, because of his experience of paper manufacture, was sent to Tottenham mills. It did not take him long to realise that certain infractions of the tax laws were being committed, and he kept a careful record of them. In August 1749 Israel Johannot died and his widow tried to carry on with the help of the foreman. Cooke now told her that the sum of the penalties incurred were double the value of the mills, and he proposed marriage as the only means of keeping quiet. As he was a handsome man, she accepted, and Cooke thus became possessed of her considerable property. When the lease of the Tottenham mills expired in 1770 he could not obtain a new one because it had already been granted to Wyburd. Cooke was furious and refused to give up the premises. Wyburd responded by taking off part of the roof; this so angered Cooke that, when he met Wyburd later, he took him by the breeches and threw him into the Lea. Cooke was sued for this and eventually left Tottenham; he took on a sugar factory near London Bridge but this enterprise soon failed owing to his lack of experience. To redeem his losses he became miserly, and gained notoriety. He would pretend to faint outside a stranger's house and, when helped by the tenants, would notice the children, making such enquiries as to lead their parents to suppose they might be remembered in his will. He made many 'friends' in this way and from gifts received of turkeys, pheasants and wine managed to cut his household expenditure, and make money by selling some. His wife died broken-hearted by his ill-treatment, while Cooke lived on in squalor and decrepitude. He left the bulk of his money to charity.

The mill was soon joined by an oil mill on the opposite (south) side of the river, which Wyburd also leased from Cooke, with the right to sublet it. This seems to have been built between 1770 and 1773; in a deed of 1770 the corn mill is mortgaged to a Charles Mesman for a loan to effect improvements, and another deed dated 1773 confirming the mortgage mentions 'the new erected oil mill'. An indenture dated 1 May 1780 between Turner of Paul's Wharf, London, sugar refiner, and Richard Grindall of Austin Friars, transferred 'all that new erected oil mill situate . . . in the parish of Tottenham . . .near to the corn mill of the said Edward Wyburd from Wyburd to Turner with assignment of mortgage to Grindall'. Shortly afterwards another paper mill was built in the area, but was short-lived. It is shown on Andrews and Drury's map of the county 65 miles around London in 1777, but there is no reference to paper-making at this site thereafter.

On 23 February 1788 both corn and oil mills were destroyed by fire but new mills, on opposite sides of the road (the oil mill on the south), were soon erected. The oil mill was at the time of the fire still in the possession of Barnard Turner. Soon after the general auction of the estates of the lord of the manor, Sir Henry Hare Townsend, the new mills were sold to John Cook, described thus in the sale catalogue of the estate on 25 September 1789:

'Lot XXX — the corn and oil mill, with outbuildings for conducting and carrying on extensive works; and also, several enclosed and common marshlands; containing in the whole about 53 acres; let to Mr Wyburd, at a ground rent of £210 p.a.'.

The property was worth more. The catalogue noted that the River Lea Navigation Act of 1779 safeguarded the flow to the mills. The lessee had the liberty to underlet or sell the oil mill during his term, discharged from all demands by the rent, the corn mill being considered sufficient to meet them.

In about 1809 the mills and lands passed to Charles Pratt, who had bought the remaining 17 years of Wyburd's lease. In 1815 the remaining six were sold to a firm named Curtoys and Matthews. The tenant received from the River Lea Company payment of £50 pa and there was a toll payable for persons, cattle etc passing over the mill bridge. In 1810 tolls were also paid by the landowners in Mitchley Marsh, who had to cross the mills' lands after the parish declined to rebuild a bridge to the marsh from Down Lane.

The mills were badly damaged by flooding in 1817 and were not at work for nearly the whole of that year. In 1824 a coal wharf was added, indicating a steam engine supplemented the waterwheels. Nathaniel Matthew obtained a 14-year lease from John Cook in 1826 at £800 pa, and about 1832 disposed of it to Edward Bell. The documentation describes it as 'all that timber-built mill containing in length from east to west 126 feet and from north to south 110 feet together with the waterwheel shaft and old waterwheel and the other wheels and machinery thereto belonging specified in the shedule hereunder . . . and also all that messuage or tenement and drying kiln, and all that workshops, stables, outhouses, and also all those 13 several parcels of land containing 24 acres 1 rood 16 perches, and all those 3 parcels of marshland containing 2 acres 2 roods 10 perches . . .'.

In 1836 John Cook sold the freehold to the New River Company, who granted a fresh lease to Edward Bell when the first expired in 1840. In that year Robinson wrote in his *History of Tottenham*, 'There are six pairs of stones in the corn mill, capable of grinding 300 quarters of wheat per week in the autumn, and in the winter months when there is an abundant supply of water; and in the summer months when the supply of water is scanty and uncertain, about 150 quarters per week may be ground on the average. This mill is chargeable with annual payment of £10.00 as a composition for small tithes to the Vicar of Tottenham'.

In 1852 the mills were again devastated by fire. At the time they were undergoing considerable alteration towards expansion. Just before four o'clock on Sunday morning a watchman noticed a cloud of dense black smoke coming from the oil mill. He immediately raised the alarm, but the fire spread rapidly due to the highly inflammable nature of both the oil and the seeds being stored prior to crushing. The sight of it was described as of 'fearful grandeur', with the flames mounting to a terrifying height and the blazing oil running over the road and into the river. Both the oil and corn mills were totally destroyed.

The buildings belonged to the New River Company; they do not seem to have been insured, and were not rebuilt. The origin of the misfortune, which threw about 50 families out of work, was never discovered. Some buildings associated with the mills, including a chimney stack, were still standing in a ruined condition in 1923.

Lace and Cotton Mill

This mill was offered for sale in February 1829, when it consisted of two lace factories, with an engine house and a steam engine of 18 horsepower. The Parliamentary Return of Mills and Factories in 1838 lists it as a cotton mill and silk mill. The former had a 50 horsepower steam engine and employed 17 people who were all juveniles; the latter had an 18 horsepower engine and employed 154 persons, of which 46 were between nine and 13 and 64 between 13 and 18.

TWICKENHAM

(1) River Crane. Early mill near Hanworth Bridge.
(2) Mill of unidentified type, standing 15th century.
(3) River Crane, Fulwell Mill. 144728
(4) River Crane, gunpowder mills. 128728

In the Middle Ages there was a mill on the Crane at a place called Oldford, ner Hanworth Bridge. It was partly in Isleworth manor (and Twickenham parish) and partly in Hanworth. It disappeared between 1340 and 1351. In 1486 Elizabeth York owned a 'mill-house with querns'.

A mill was built by Thomas Betts on a reinforced stretch of the Crane at Fulwell, where Mill Road meets the southern stream of the river by the former Fulwell Park, some time before 1753. Then it was referred to as the New Mill, suggesting it replaced an older structure. By 1767, after use as a copper mill, when it had a low breast shot wheel installed by John Smeaton, the engineer who built the Eddystone lighthouse, it was crushing seeds to produce oil. The owner in the Poor Rate Book was John Merchant or Marchant who operated the 'New Oil Mill' with his brothers, Geoffrey and George. Geoffrey was the owner in 1770 and by 1771 the Marchants had been succeeded by Charles Barrow. In 1784 both house and mill passed to Addis of Addis & Winslow, and in 1785 Thomas Winslow purchased the estate from his partner, along with the surrounding land, Warren House and Fulwell Lodge, for £7,800. All these owners tended to live at Fulwell Lodge and own land on which they grew crops, possibly processed at the mill. Under Barrow and Winslow, and especially Winslow, under whom it expanded, as the 1791 employment of a clerk indicates, the business prospered. Ironside writes in his *Memoirs Of Twickenham* of great quantities of linseed oil sent to London daily. By Winslow's time the mill also manufactured cattle cakes. By 1799 part of the building was used to dry tobacco, while another was a windmill. After Winslow's death in 1805, history repeated itself when his business partner

George Thackrah purchased the mill, the Lodge being sold separately. Thackrah also bought the house, office, and the Warren Farm containing 48 acres, for £4,600. He was not popular; in 1820 and 1832 he was involved in litigation because he closed the right of way which led from the side of the mills across the fields to Whitton. He soon departed and the business was carried on by Messrs Allan and Pitcairn, who remained until the mid-1840s, when the mill was converted to manufacture paper.

Then the owner was Timothy Healey, described as 'paper, millboard and pasteboard maker. Not long after this it passed to William Brindley and Thomas Boyle, who used it to produce papier maché. Their partnership was dissolved in 1850 after which William Brindley ran it on his own. In 1851 the mill had two beating machines. Around 1867 the paper-making business ceased, and during the 1870s the mill was badly damaged by fire. A photograph taken c1890 shows it as a roofless but impressive derelict. The waterwheel was still in palce when the photograph was taken, and survived the mill by several years; it was there in 1900, but all traces of the building have gone.

Twickenham is most famous for the Hounslow Gunpowder Mills (as they tended to be known, although the site actually lay within Twickenham parish). Like so much of the land in this area the site was originally leased from the Duke of Northumberland.

The mills were founded in the 16th century and functioned until 1926. In 1821 Curtis and Harvey took out a 12-year lease, renewed on its expiry; in 1871 they acquired the freehold from the Duke. By 1850 steam power had superseded the waterwheels. The remains on site are less substantial than those at Bedfont, with the exception of a possible windmill tower, which certainly had a part to play in the functioning of the works; they consist of a number of millstones, a few fragments of brickwork here and there, and the baffle mounds constructed near the works buildings to contain the blast from any explosion. As with the Bedfont remains, the site lies within pleasant surrounding, in this case those of Crane Park, which contains a nature reserve.

Various explosions took place at the works though some occurred long after the mills stopped using water power. It appears the baffle mounds were ineffective in reducing damage. By no means all were recorded, particularly if they did not result in fatalities. Some certainly took place in 1758, 1772, particularly between 1796 and 1813 (probably due to the increased activity during the Napoleonic Wars), 1826, 1839, 1842, 1850, 1859 and 1915. They tended to be heard, and the shock felt, for quite a distance. That of 1772 was audible in Weybridge and Walton-on-Thames, about five or six miles away, while the writer Horace Walpole recorded in his diaries that reverberations from the blast had shattered two marble busts in the entrance hall of his house at Strawberry Hill. In 1850 shock waves were experienced in Brighton and Hove. The effects on the bodies of those killed were often gruesome. Cavalry from Hounslow Barracks were sometimes employed in the unsavoury task of collecting the severed heads and limbs and mangled torsos.

Twelve people died in 1796, two in 1839 and in 1842, seven in 1850 and 1859. By the late 1850s there was increasing concern, among both the workforce and the general public, about standards of safety. At least one strike is recorded and one 'senior and very experienced' foreman resigned in 1858 over the installation of a new press, which he felt to be dangerous and which in fact played a major role in the following year's disaster. After the 1859 tragedy new safety precautions were introduced, and there were no really serious incidents apart from that of 1915, in which no-one died.

In 1675 John Davis of Bedfont was charged with 'threatening to throw a squib to blow uppe [the] powder mill' at Hounslow.

UXBRIDGE

(1-3) 13th century mills.
(4-6) 14th century mills.
(7) Mede mill. Standing 14th century.
(8) 17th century mill.
(9) Crouch Mill. Standing 1649.
(10) River Colne. Standing 1746.
(11) Frays River. Town Mill. Standing 1636. Ancestor of Fountains Mill. 053844
(12) New Mill.
(13) Frays River. Fountains Mill. 053844
(14) River Colne. Upper Colham Mill. 047836

In 1265 Godfrey de Heddesore owned three watermills in Uxbridge. In the 14th century the lord of the manor of Colham had two mills under one roof in the town and two separate ones south of it. A mill called Mede Mill, also belonging to the lord of Colham, existed by 1409; it was then horse-powered but functioned as a watermill from 1419, in which year a wharf was constructed. In 1537 permission was granted to install an additional mill for grinding malt. In the 1530s there were two mills at Uxbridge, one driven by the Frays River and the other by the Colne, near the Oxford Road at the west end of the town. In the 17th century a watermill in the area belonged to Stanwell Manor. In the 18th century one or more of the Uxbridge mills was used for paper-making. Town (later Frays or Mercer's Mill) on the Frays at the west end of Uxbridge and Crouch Mill, presumably identifiable with two 14th century mills which had the same names, were again mentioned in 1636 and 1649. In 1648 Crouch Mill and the adjacent mill house were leased to Samuel Bonsey, a London merchant. Before 1746 another mill, on the Colne west of the town, had apparently been built.

The mills attacted the attention of early travellers. Leland in 1540 describes the two bridges from which the town gets its name, 'one over the great arme of Colne River while the lesser arm (Frays River) goeth under the other bridge, and each of them serve a great mill . . .'.

Three mills, the New Mill, Town or Fountains Mill, and Upper Colham mill, survived into the present century.

New Mill

This was erected in 1836, the building being the work of J. Shoppee and the millwrighting, of J. Penn. It functioned as a stone mill until 1902 when converted to a roller mill, the change increasing the mill's output to six sacks per hour from two. From 1890 it was run by William King. Kings Ltd operated the mill in 1902 but some time later it went out of use.

Fountains Mill

This undistinguished-looking brick building, which still stands at the western end of the High Street, is the descendant of a mill which existed in 1767. This was of brick and timber with a tiled roof and was occupied by the Mercers, a well-known milling family. In those days it was known as the Great Mill, suggesting a building of considerable size. In 1782 the mill was run by John and Nicholas Mercer, who insured it against fire for £700. It was in fact burnt down in 1796, but rebuilt. In 1805 a Commisson of Bankruptcy was awarded against John and Nicholas, yet the business seems to have continued and the mill remained in the family until taken over by E. & J. Fountain, from whom the name by which it is still known is derived, shortly before 1895.

In 1909 the mill was driven by a 60-80 horsepower compound condensing steam engine assisted by a 40 horsepower Action turbine. The waterwheel, which was a large affair enclosed within the building, had by now been relegated to driving just an elevator and the provender plant. By now a roller plant, built by Simons of Manchester, had been installed, although the stones remained and were still used for wholemeal flour production in 1956. Water continued to be employed in some small capacity at the mill until quite a late stage, although probably *via* the turbine only.

The mill was badly damaged by fire in 1954, and recorded in a series of photographs of the incident, now at Uxbridge Library; however, it continued to function for a couple more years at least. Since ceasing as a mill it has served the community in several different and important ways; until 1981 it was the home of an Industrial Training Unit, and since 1985 has been a Youth Centre and Substance Abuse unit.

Upper Colham Mill

This mill, which stood on Uxbridge Moor, off the road to Iver, is probably identifiable with that shown on a map of 1746 as on the Colne west of Uxbridge. In 1832-4, Spencer Homewood was the miller. His successors were Robert Tate Junior (1851), D. Russell (1862, 1866), and finally William Edwin Hamaton (1895, 1899) who also worked the mill at Denham in Buckinghamshire. The mill passed in 1919 to the Bell Punch Company, who were the owners in 1942. It did not survive much longer.

WESTMINSTER

(1) Tyburn. Early mill, Westminster Abbey. 302793
(2) 18th century mill. 303792

In the 12th century one of the rules of Westminster Abbey, which then belonged to the Benedictine order of monks, was that there should always be a mill; the principle was that the monks should be as independent as possible. It was Abbot Litlington who built the 'molendinus aquaticum', as documents in Latin describe it, driven by the Tyburn, between 1362 and 1385. It is from this mill that the name 'Millbank' originates. The mill was not operated by the monks themselves, but they took a certain share of the rent. The Abbot, like secular magnates, was allowed to have his corn ground free.

The mill is recorded in 1557, when on the night of 22 January one John Harris drowned while drunk, after falling into the dam in which the water was stored, and in 1644, when 11 shillings were paid to John Redwood for 'charges upon sundrie indictments touching the bridge at the watermill'. Kipp does not show it in his view of this part of London in 1724, so perhaps it had disappeared by then.

Norden's map of 17th century London indicates a second mill in the vicinity, at the meeting of Abingdon Street and Millbank.

WILLESDEN

River Brent.
(1) 14th century mill.
(2) Possible 19th century mill near Parish Church.

A mill which formed part of a grant in 1325 from Richard of Cornhill to John, Vicar of Willesden, was probably a watermill on the Brent.

The Grange Museum at Neasden has a slide of a watercolour, dated 1824, which depicts a watermill near Willesden Parish Church. It is a small building with a thatched roof. As there is no other evidence after the 14th century it is possibly a figment of the artist's imagination.

YIEWSLEY

(1) River Colne. Kelsey Mill. Burnt down 1873. 049816
(2) Huntsmoor Weir. 048815

This mill stood at the rear of Huntsmoor Park. In 1816 it appears on the Ordnance Survey Map. It was known either as New Mill or Kelsey Mill. Henry Kelsey was miller in 1862 and 1870. It was destroyed by fire in January 1873. The mill house still stands today.

A single-storey remnant of a building, within which is an elderly Jonval turbine, survives. The mill was grinding, using electricity in the late 1930s. The mill is almost certainly identical with Kelsey mill, Yiewsley, indicating a reconstruction following the fire of 1873.

Finally, there were once waterwheels between the arches of London Bridge. As they were on the northern end of the bridge, they should be regarded as in Middlesex. They were destroyed in the Great Fire of 1666.

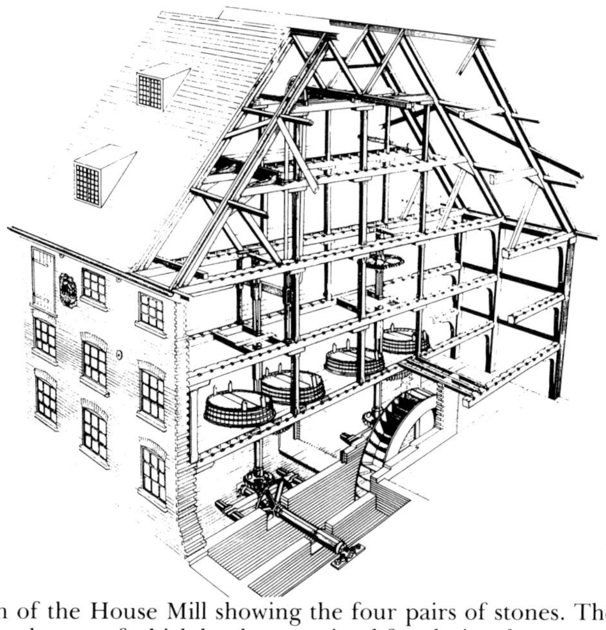

Diagram of the House Mill showing the four pairs of stones. The two waterwheels, one of which has been omitted for clarity, drove two pairs each *via* a spur wheel and stone nuts. (JHA)

ABOVE: Brentford watermill in 1903. (HLS, CLLHC) BELOW: Colham Mill, c1805. (GL, COL)

ABOVE: Bedfont gunpowder mills in 1883. BELOW: Rabb's Mill, Cowley, late 19th century. (HHS)

ABOVE: Fountains Mill, 1995. (GB) BELOW: Harefield Copper Mills, 1995. (GB)

ABOVE: The east side of the Old Mill, Old Mill Lane, Cowley, Uxbridge, shows the unusual arrangement in which the weatherboarding is outside the frame. The arch on the wall marks the position of an earlier waterwheel. (JKM) BELOW: The downstream view shows the three original waterways. These are still in use. (JKM)

ABOVE: Lea Bridge Mills and mill house, Hackney. (HAD) BELOW: The Calico Mills at Isleworth, 19th century. (By permission of the British Library)

ABOVE: The Good Intent Mill, Isleworth (GL, COL) BELOW: Kidd's Mill, Isleworth, in its prime. (LBH)

ABOVE: Upper Mill, Stanwell, 1994. (GB) BELOW: Tottenham Mill, date unknown. (GL, COL)

Twickenham: Fulwell Mill in decay, c1890. (LBRUT)

LEFT: Stone dressing at Fountains Mill, Uxbridge, 1956. 'Dressing' was the task of trimming the face of the millstone, using tools called bills like that shown here, so as to maintain a smooth and even surface for grinding. (UG) RIGHT: Black Jack's Mill Restaurant, 1995. The mill race still flows under the building. (GB)
BELOW: Steam wagon at Fountains Mill in an earlier era. (HHS)

Windmills

The development of the windmill in Middlesex followed a similar pattern as elsewhere in England. Windmills first appeared in Europe during the 12th century; the earliest Middlesex reference, at Harrow, occurs in 1236. A mill may have existed at Clerkenwell in 1134, but the evidence is inconclusive.

Mediaeval windmiils were mostly 'post' mills — where the boxlike main structure is mounted on bearings on top of a post which serves as a pivot, around which it can be turned by means of a beam called a tailpole, projecting from the rear of the structure through the access ladder, to keep the sails facing into the wind. The post was surrounded and supported by a trestle structure. This was originally exposed but, from the mid-18th century (if not earlier; see the entry on Southall windmills), enclosed within a circular building called a roundhouse. This protected it from the elements and served as extra storage space for grain. The other two main types, the smock and the tower mill, did not appear until later. In both, a stationary tower (of wood in a smock mill, so called because of a fancied resemblancc to thc garment once worn by rural labourers, and of brick or stone in a tower mill) contains the machinery and millstones. Only the roof section or 'cap', whose base supports the axle or windshaft that carries the sails, is rotatable.

Contemporary carvings and manuscripts suggest the early post mills were small, with room for only one pair of millstones and one person at a time. Later post mills might have two or three pairs, one in the head and one in the tail, or two side-by-side in the front or breast with perhaps a third in the tail. Being constructed of wood, they were fragile. Like watermills, they were initially manorial mills, owned by the lord and operated by a miller, usually a tenant.

As a technical novelty, largely untried, the windmill was regarded with suspicion. It was seen as a supplement to the watermill, rather than a replacement. It tended to be built in areas where there was already an established watermill, so that, if the watermill was out of action the windmill could take over. Unfortunately this created problems — the mill might not be used, or maintained, so might often deteriorate, especially as it was relatively fragile and, easily damaged by the elements. The chequered career of that at Ealing reflects this vulnerability. Another fate which threatened Middlesex windmills was demolition by the poor for firewood. This seemed likely at Pinner, largely because the mill was falling to pieces anyway.

These factors made upkeep costly, which deterred prospective patrons. Repairs were often postponed or not carried out at all. Perhaps the lord simply grew tired of a troublesome responsibility and simply let the mill fall apart.

A mill's survival could also be jeopardised if it was built in hard times, like that at Upper Halliford, Shepperton, in the 1380s. However, specific local circumstances at Harrow gave the windmill an advantage, and it held its own against the watermill.

In time, the windmill became an accepted part of the rural economy. The crucial factor ensuring its ultimate success was population growth, which increased consumption; in the absence of water power, a windmill was the answer.

Nevertheless in Middlesex the windmill was less common or successful as the watermill. Over 100 windmills existed in the county between the 12th century and the 20th, but the total at any one time was not more than 30. Watermilling dominated because of the Thames, Lea and Brent. The distribution of windmill sites throughout the county was fairly even; Middlesex is generally low-lying, and thus excellent terrain for windmills (which are not normally built on hills, because this means more exposure, and therefore more repairs). There were fewer restrictions on location than for watermills, which could only be built by water.

The Industrial Revolution saw windmills undergo considerable, if gradual, technological change. Tower and smock mills became common and some were much larger than many post mills, Brentford being a good example, with room for two, three or even four pairs of stones. Wooden components were often replaced with iron. Common sails — in which the wooden framework was covered with canvas which performed a similar function to the sail of a ship — were replaced with spring or patent sails. Here a number of shutters were mounted in the framework, opened or closed as necessary to lessen wind resistance, by means of spring in the former and a system of adjustable weights in the latter. Many mills, like Enfield and Willesden (Shoot-Up Hill), had one pair of common sails and one pair of springs or patents as a cost-saving exercise.

Instead of being turned by hand *via* a tailpole or a rope/chain — a laborious task — the caps of smock and tower mills were often winded by a fantail, patented by Edmund Lee in the 1740s and resembling a miniature windmill. When the wind changed, it struck the fantail at an angle, causing it to rotate; through gears meshing with a toothed rack on top of the tower, it turned the cap and sails to face the wind. The fantail was usually behind the cap, in place of the manual gear. A few post mills also had fantails, usually mounted on the end of the tailpole or access ladder.

In a windmill the drive to the stones is from above — the reverse of the operation in watermills. Arrangement of the drive to the stones differed between post mills and smock/tower mills. The former could be one of three types: (1) One pair of stones in the front or breast of the mill was driven from a brake wheel — so called because a brake could be applied to it to stop the mill — mounted on the windshaft, and a second pair in the rear, from a tail wheel; (2) two pairs of stones side-by-side in the breast were driven from the brake wheel *via* a wallower and great spur wheel, and a third inthe tail by the tail wheel *via* a stone nut — this method was a comparatiely late development; (3) two pairs only were positioned in the breast and driven from the brakewheel *via* wallower and spur wheel, with no tail wheel.

In smock and tower mills there was usually a wallower, upright shaft and great spur wheel driving the stones through stone nuts. In a post mill ancillary equipment was driven *via* bevel gears from the brakewheel or tailwheel, in a smock or tower mill from a crownwheel on the upright shaft, again *via* a layshaft and bevel gears.

Although the first large steam-powered flour mills appeared near the end of the 18th century, in most of England windmills and watermills were built for another 100 years, some still working well into the 20th century. The comparatively early decline of Middlesex windmills was due to London's growth. The capital absorbed parts of Essex and Kent in the East, Surrey in the south, Hertfordshire in the north and Middlesex in the west. Middlesex bore the brunt of this expansion, ceasing to be a separate entity. An entire county, largely rural, disappeared under bricks and concrete and lost its identity. There has been nothing comparable to it in English administrative or social history. Most of what is now London north of the Thames, including the seat of government at Westminster, was once Middlesex.

As farmland was sold off and the rural working class moved to urban jobs, the way of life of which the windmill was essentially a part disappeared. It was swept away by development. It did not disappear entirely; although windmills for grinding corn were a thing of the past, there appeared an assortment of mills for other purposes, some erected on the roofs of other buildings. As late as 1892 a windmill was erected on a building in the City Road, East London, to drive machinery for flour production.

By 1850 there remained at the most half-a-dozen mills, most of them in rural areas. These too succumbed, either to urbanisation or new sources of power such as steam, oil and gas. Hounslow and Edmonton mills worked latterly in partnership with, or as supplements to, steam-powered mills, which helped to prolong their useful lives. The last mills to cease grinding were those at Enfield and Edmonton; the former may well have carried on into the early 1900s.

The number of new windmills in the county varied over 800 years was: 13th century — 7; 14th century — 14; 15th century 1; 16th century — 6; 17th century — 26; 18th century (up to 1780) — 17; 19th century — 25; 20th century — 2.

The total number of recorded sites is 100.

Windmills tended to differ from one area to another but there is little evidence to define the 'typical' Middlesex windmill — if there ever was such a thing.

Photographs offer the best clues; those of six windmills survive, excluding contemporary ones of the windpump base at Clerkenwell and the possible windmill tower in Crane Park, Twickenham, and there are reliable non-photographic illustrations of several others. The photographs are of Edmonton, Enfield, Hampton, Hounslow, Monken Hadley and Willesden mills. That of Hampton has been heavily retouched, while Monken Hadley is little more than an outline.

The majority of windmills in Middlesex disappeared before the advent of photography in the 1840s and 1850s, but those which remained were in the border areas of the county, where regional differences were blurred. Edmonton and Enfield mills, for example, had certain features also found in the Essex/Bedfordshire/Hertfordshire region.

There is one feature of the smock mills which may have been characteristic. Four — Acton, Enfield, Hampton and Sunbury — had little or no brickwork beneath the wooden tower. It was usually the custom for smock mills to rest on a brick or stone base of at least one storey.

Middlesex Windmills 1100-1940. (PB)

WINDMILL GAZETTEER

ACTON

(1) C17 mill, Acton Park. 208803
(2) C18 mill, Twyford Avenue. 195805
(3) Smock mill at or near site of (1). 208803
(4) Possible mill, Mill Hill Park. 198797

The first reference at Acton is in 1622, when a document mentions the waste in front of the mill. It stood at the eastern end of Church Field, on land now part of Acton Park. Later, there was a mill on or near the site of the present Elms public house in Twyford Avenue. This is shown on a map of Acton dated 1700, which is preserved at Gunnersbury Park Museum, and on Seller's map of Middlesex in 1742.

Some time in the 18th or early 19th century a smock mill was built on or near the site of the mill standing in 1622. In 1801 it was held, along with the mill house and some of the surrounding land, by Robert Nunn. Isaac Adams was both owner and miller in 1823-4. By 1841 it was the property of Ann Adams who also had a smallholding of about 10 acres, including an orchard. She employed Joseph Franklin as miller. Ten years later it was operated by William Harding, who lived in the mill house with his wife and two young daughters. Latterly the mill was leased from a wealthy local family named Billington.

The mill is shown on the Enclosure Commissioners' Map of 1859, but was absent from the 1865 Ordnance Survey map. Surviving traces were probably obliterated around 1877 when the Goldsmith's Company, who had acquired the land, began to build.

However, it features in several watercolours, the best by John Varley, dated 1835. It stood on a shallow brick base without a stage and had a hemispherical cap, winded by a lofty fantail, and four common sails. A painting by J. Gudet shows how two artists can represent the same mill differently. Gudet makes the sides of the smock tower more or less vertical, whereas most smock mills had tapered sides, and the sails project much further forward than is likely.

The name Mill Hill Park, known in 1810 as Windmill Hill, suggests the existence of a fourth Acton windmill.

ASHFORD

Early mill.

A mill is recorded here in 1277 and subsequently it was out of use by 1309, when it was described as 'broken' and without a millstone.

BLACKFRIARS

Standing 1696-1748. 324808 (approx.)

A number of illustrations, the earliest of which dates from 1696 ('A Long View of London and Westminster taken from Southwark', by Philip Lea and Robert Morden) show a windmill at Queenhithe. Views of 1720 and 1748 depict it standing on a tall building, around the top of which is a gallery, while others from 1710 and 1724 are similar. Another view by Hoffner represents it strangely as four sails mounted at the top of a tall mast standing on the river bank.

LEFT: Windmill at Queenhithe, 1696. (PB) RIGHT: Queenhithe windmill, by Kipp, 1748. (PB)

BOW

(1) Old Ford New Town. Standing 1844. 365835
(2) Near Wrexham Road. Standing 1844. 375832

Two Bow windmills are shown on the 1844 Ordnance Survey map. The first stood in Old Ford New Town, somewhere near the top of the present St Stephens Road and between Roman Road and the Old Ford Road. The second was in the vicinity of Wrexham Road, possible on the site of the later trolley 'bus depôt.

Directories list J. S. Ward (1823-4), Edward Long (1845), William Cooke (1848) and W. Bonwick (1851-2) as millers under Bow, but it is not known which of the two mills they worked.

The following advertisement appeared in the *Chelmsford Chronicle* of 30 March 1838: 'WINDMILL TO BE SOLD. A powerful post windmill with one pair of very good 5ft French Burr millstones, cast iron windshaft and cross, hung on ditto, for sails, a very good 9ft 6in. brakewheel and brake; 4 good sails, from centre to point 35 ft 8 inches. The whole with roundhouse, post etc. in good working condition and will be sold a bargain. Further particulars apply to Messrs. Hunter and English, millwrights, Bow, Middx.'.

This mill may not have been at Bow but an Essex mill, perhaps one of those at Stratford, for a millwrighting firm's catchment area often transcended county boundaries.

In a contemporary sketch of the demoliton of Bow Bridge in 1835, a tower or smock mill with a domed cap appears in the background.

BRENTFORD

(1) Early mill, near Boston Manor Road. Gone during 18th century. 172786 (approx.)
(2) Smock mill, Old Brentford. Stood by river. Probably gone during first half of 19th century. 186779

A windmill was standing in New Brentford, near Boston Manor Road, in 1698 and 1703, and had evidently been in existence for some time, for by 1670 the field in which it stood, formerly New Brentford Field, had been renamed Windmill Field. The mill gave its name to Windmill Road in what is now Little Ealing. A late 18th century legal document concerning a boundary dispute sought to establish the site of the by then vanished mill as a reference point; it states that the mill had been gone for too long to prove its position. Martin Mason asserted in the 1950s that it survived into the late 19th century and was featured on a playing card in the 'Happy Families' series; there is no other evidence for a mill here at such a late date and no card survives. Mr Mason added that the picture seems to place the mill just below the gate of the open space known as the Ride.

By the late 18th century Brentford had become an important centre for flour milling; its mills were mostly water-driven, the Thames and Brent providing abundant power, but a windmill was built on the south side of the road beside the Thames, on the site of the gasworks. It first appears on a 1777 plan of Ealing parish. A pencil sketch by S. H., Grimm dated c1780 shows a large smock, with a stage braced to a substantial brick base — a feature which distinguishes it markedly from most other Middlesex smock mills — by diagonal supports. The hemispherical cap is winded by hand and there are four common sails, on which the canvas is shown furled, indicating the mill is at rest. The squarish structure beneath the stage, from which appear to hang a number of ropes or chains, was probably a lucam such as is more usually found on watermills.

The mill probably did not remain in use long enough to receive modifications such as a fantail and shuttered sails.

In 1784 it was owned by Joseph Shaw, a maltster, and tenanted by David Roberts. By 1789 Roberts owned it with Thomas Smith and Thomas Harrington as tenants. Later it was operated jointly by Richard Kidd and Richard Bax but the partnership was dissolved in March 1801, after which James Kidd ran it on his own. It had ceased work by 1828; a view of the Thames from London to Richmond published that year shows it without sails, noting its use as a basket factory.

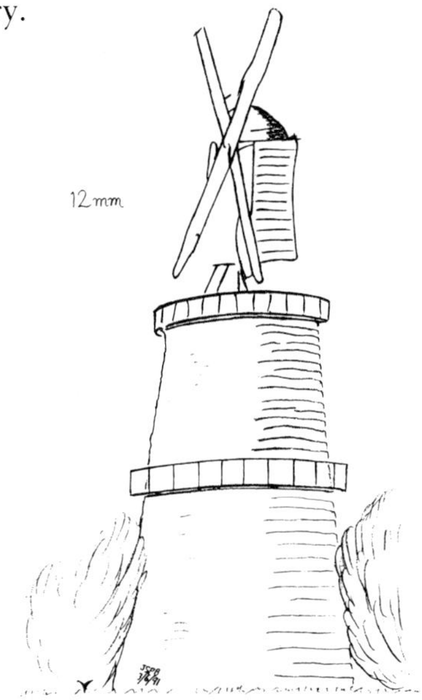

LEFT: The smock mill at Brentford, from a pencil sketch by S. H. Grimm, late 18th century. (SB) RIGHT: Windpump at Chelsea Waterworks, from an 18th century print by Antoine Benoist. (SB)

A post mill features in a geographical card game, common in the second half of the 19th century, entitled 'Counties of England'. The card concerned depicts the industries of Brentford. It has a roundhouse, sails of indeterminate type and a ladder braced to the body of the mill by a rope or chain. It stands on a hill with several industrial buildings, whose chimneys belch forth smoke, in the background. There is no indication that it represents an actual mill, and it was probably included as a symbol of flour milling in the local economy.

BROMLEY BY BOW

C18 mill. 379825 (approx.)

Rocque shows a windmill here on his map of 1754.

CHELSEA

(1) Tothill Fields, smock mill. Possibly same as (4)
(2) Tothill Fields, post mill.
(3) Smock mill at 287780.
(4) Smock mill at or near 287780.
(5) Chelsea Waterworks, pumping mill.
(6) Various other apparent windmills, probably identifiable with nos. 1-4.

More than one windmill existed in the Chelsea area around the beginning of the 19th century. There are a number of pencil sketches which show a smock or tower mill at Tothill Fields, near Willow Walk (the present Warwick Way). The earliest is by W. Capon and dated 1799. Another, also by Capon, shows the mill with a dome cap with a decorative finial, winded by a chain and wheel, and common sails. In 1806 Capon sketched a post mill in the area. A note reads 'Where the horses are killed' and that the view is of Tothill Fields from Marsham Street, looking west.

Farriers and Mason in *Windmills Of Surrey And London* thought it 'possible, but unlikely, that [the smock mill mentioned above]' is the same as that shown in a panorama of the Thames dated 1828 . . . to the east of the waterworks within the premises of a lead factory works opposite the eastern end of Battersea Fields'. This mill also features in a drawing by Augustin Rischgitz dated 1815 and now held at the Chelsea Library, along with another which stands among trees at the point where the Grosvenor Canal joins the river.

Curiously, none of the above mills are on the various contemporary maps. They could have been the product of artistic licence, but that all of them should be ommitted from maps seems rather odd. With one exception the mills had probably disappeared by 1830, when the area was being rapidly developed. The surviving mill was standing in 1832 in an area north of the river known as Thames Bank. A newspaper cutting, now in Chelsea Library, includes the following:

'Thames Bank Windmill: The Inhabitants of Chelsea and its vicinity are respectfully informed they may be supplied with BREAD & FLOUR of the very best quality and free from any Adulteration, at the lowest possible Price, for Ready Money at the above Mill.

MALT, HOPS, CORN, BIRDSEEDS E&c.'.

The Library also holds a watercolour of a white smock mill at Chelsea Reach; this is of stout proportions which resemble Upminster mill in Essex. The cap is a dome with a ball at its top and a railed wooden structure surrounds the base. An engraving entitled 'Don Salter's Walk', from a painting by Cecil Lawson, shows two windmills of smock or tower type near the river. Both have fantails, that of the nearer mounted above the extension for the former manual winding gear. This mill also has a guide pole for the striking chain of the patent sails, and the cap is a dome. It bears a strong resemblance to the mill sketched by de Court at Westminster. An engraving of the Royal Hospital based on Kipp's view from the north-west shows two windmills in the background. There are a number of other illustrations which show windmills in small detail.

A 1752 view of Chelsea Waterworks by J. Boydell includes what appears to be a tower mill without cap or sails. There is no evidence that a windmill ever existed at the works, and it is more likely to have been a water storage tower. However, an illustration in the British Museum may shed some light. It shows an odd-looking structure consisting of the body of a post mill mounted on a substantial brick tower with which it is small and narrow by comparison. This tower has two stages, one about halfway up and another at the top, which presumably serviced the sails. What this curious mill was used for we can only speculate, but its proximity to the waterworks suggests it was probably for pumping water. It is conceivable that the tower shown in views of the works could have been the brick base of this mill with the post mill superstructure removed.

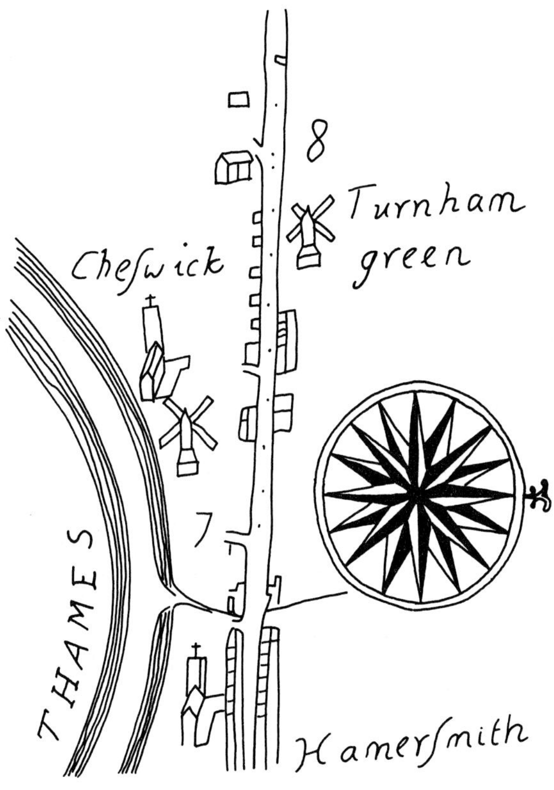

The two windmills at Chiswick; detail from Ogilby's *Britannia*, 1675. (PB)

CHISWICK

(1) North of High Road. Standing 1675. 211786
(2) Between High Road and river. Standing 1675. 224784 (approx.)

Two windmills are shown at Chiswick in Ogilby's *Britannia*, a road map of the British Isles published in 1675. One is to the north of the High Road, and probably that which stood with a cottage on the waste of the prebend manor at Turnham Green c1650.

The second is to the east of the road near the river. Both mills have vanished without trace, but the former is commemmorated by Windmill Road and Windmill Passage, and by the Windmill public house (whose sign depicts a Dutch drainage mill from which it was probably far removed). There, whoever has been searching in vain for some remnant of a Chiswick windmill may drown his sorrows.

CITY OF LONDON

Cripplegate.

In 1598 a windmill and a tenement on land known as Mountmill, in the parish of St Giles at Cripplegate, featured in the will of Richard Devenish. In a related legal case the mill was 'hidden by the City of London'. Three years later, in 1601, it came to grief in a storm. It was probably never rebuilt, and there is no evidence of a later mill. However, a windmill at Rusper in Sussex, now no longer standing, was known as Cripplegate Mill, allegedly moved from Cripplegate in 1810. The name was later given to a smock mill at Southwater, near Horsham in the same county.

CLERKENWELL

(1) Possible early mill.
(2) Early mill. 315827
(3) Early mill. 317825
(4) C16 mill.
(5) Pumping mill, New River Head. Base remains today. 312827
(6) Early mill, East Smithfield.
(7) C16 mill. 319826
(8) Finsbury Fields, near Moorgate. Six mills altogether in C16 and C17. 327922.

If Professor E. J. Kealey is right, Clerkenwell may have possessed a windmill as early as the beginning of the twelfth century. In his *Harvesting The Air* he writes:
'About 1144 Jordan Bricett established a Priory for Augustinian Canonesses at Clerkenwell . . . and granted to the nuns and to Robert their chaplain fourteen acres and a "place and site for making a mill" (locum et sedem ad molendinum facendum). Neither the extant copy of the Charter recording Jordan's grant nor the copies of its confirmations refer specifically to a windmill, but . . . the word "molendinum" was sometimes used interchangeably to mean windmill and watermill. The sixteenth-century antiquarian John Stow stated explicitly that in 1100 Jordan Bricett gave the canonesses a site upon which to erect a windmill. This Elizabethan was wrong about the foundation date of the Priory, but he could easily have had access to documents specifying that Jordan's plan was to build a post mill.'

The evidence is inconclusive. Stow gives an incorrect date for the Priory's foundation and could be wrong about other things, including the windmill; given the general lack of evidence of windmills at the time, it is regrettably unlikely. Various old deeds mention a mill-stream, which points to a watermill. The weir by means of which the water was diverted to the mill was actually discovered in the 19th century.

If a windmill did exist in Clerkenwell in 1134, it would be the earliest recorded, not only in Middlesex but in Britain.

There is no doubt that the nunnery later owned windmills in the neighbourhood. A map showing the Charterhouse water supply marks two, one in the Commandary Martels and one on the other side of the road from Islington (then a country village) to London, in a field called Farncroft or the Nun's Field. The latter was built c1358 on land demised for the purpose by the Prioress for 99 years at a rent of 12d a year. The lessee was Roger de Stowe, which suggests the mill was the same as that called variously Stowell mill, Stowelymyll or Stowesmylle, described in a document of 1399 as five perches from Farncroft and beside the road from Edmonton to Westminster.

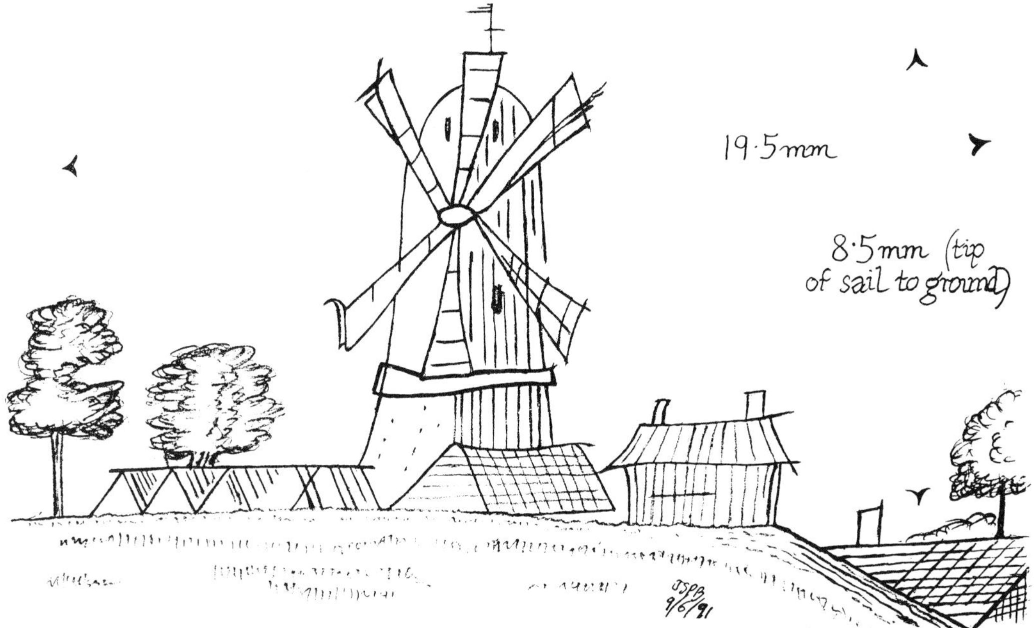

The New River Head windpump, Clerkenwell, from an early 18th century print by Sutton Nicholls. (SB)

In the 16th century a third windmill stood to the north-west of the nunnery.

The appearance of two of the above mills on a map of the nunnery's water supply has led some to conclude that they were used for pumping water. That is unlikely since such mills were a later invention. In the 18th century a pumping windmill did exist in Clerkenwell, at the New River Head waterworks, and its remains are still there today. It was built by George Sorrocold in 1708-9 to pump water to the highest level of the Upper Pond (now the Claremont Square Reservoir). It was one of the first tower mills built in England, and a print of c1712 shows that it had six sails — a rarity, and especially so in the early eighteenth century. Generally mills with more than four sails did not appear until well after 1750.

Given its comparative technical novely, it is regrettable that the structure's life as a windmill was short. An exceptionally severe gale on 20 November 1720 blew off the cap and sails, after which a horse was installed in the base of the tower to drive the machinery. Gradually, it was reduced to the stump we see today. Mid-18th century prints show the tower standing to full height and still possessing its ogee cap. The base of the mill is enclosed within a circular structure whose roof serves as a stage, and the windows are round with a resemblance to ships' portholes. An annotated view dated 5 August 1730 describes the buildings as 'The Mill from whence the Water is forced up by Horses to the House which supplies the New Pond'. By 1775 the cap had been removed and the tower reduced in height and given a battlemented top.

The portion surviving today stands 8 ft 6 inches high. It is of red brick — English bond, hand-made with alternative courses of 'headers' (*ie* bricks laid end on) and 'stretchers' (laid side on). The irregularity of the batter, which is mostly steep and almost vertical, but on one side tapers noticeably towards the eaves of the present conical roof, marks this out as an early tower mill. Millwrights accustomed to building wooden post (and occasionally smock) mills had not yet developed the skill with bricks which characterises 19th-century tower mills. The lower half of one of the windows in the 18th century illustrations is now blocked up; it is oval rather than circular, as the prints suggest.

Its survival is miraculous, considering the high property and site values in Central London, and one would not like to predict its chances of survival. It has been much knocked about and altered, as areas of modern brickwork testify. Early this century it survived a collision with a van, after which a bust of Sir Hugh Myddleton, 18th century founder of the New River Head scheme, accidentally broken at the same time, is said to have been carefully buried in the brickwork to remove the evidence from the scene of the crime.

The New River Head site is now owned by Thames Water. What remains of the mill is insubstantial and visually unremarkable. Its status as a Listed Building is understandable since it is probably the only surviving remnant of a windmill in Middlesex.

A windmill also stood in the district which came to be known as Mountmill, at the west end of what is now Seward Street. It was erected early in the 16th century on an unusually large mound.

The mill was later blown down during a furious gale and in its place Catherine of Aragon, the first of Henry VIII's six wives, built a chapel which she named the Mount of Calvary. At the end of Henry's reign, in 1547, this was pulled down and another windmill erected. The new mill is shown on Ralph Agas' map of London, surveyed in 1562-3, as a conventional open-trestle post mill of the period. It was owned by Sir George Barne and left in his will as 'My Windemyll sett and being upon the Mounte beyond Saint Johnes'.

In 1643, during the Civil War, this or its successor was incorporated into the fortifications built around London by the Parliamentarians. A broadsheet of 18 August 1643, commemmorating the failure of a 'Treacherous And Bloody Plot' by Royalists to seize the capital, in which the fortifications figured, depicts the mill without sails or rear steps and standing within a walled enclosure. There are two decorative finials to the roof ridge, one on each gable; such features were often found on 17th or early 18th century post mills — the Millwall mills had them, as does a surviving example at Outwood in Surrey. The fortifications were never actually used and the mill probably survived the war. Faithorne's map of 1658 depicts it but this was surveyed between 1643 and 1647. The site had been built upon by c1680, obliterating all traces of the mill, but the name Mountmill still survives as a nearby road.

A painting by Hogarth shows a tower mill in Moorfields, about three-quarters of a mile to the east of Peerless Pool, a popular spot for bathing and skating, in the late 18th century. The mill has four common sails and is similar to the windpump at Clerkenwell; it may have been used for the same purpose.

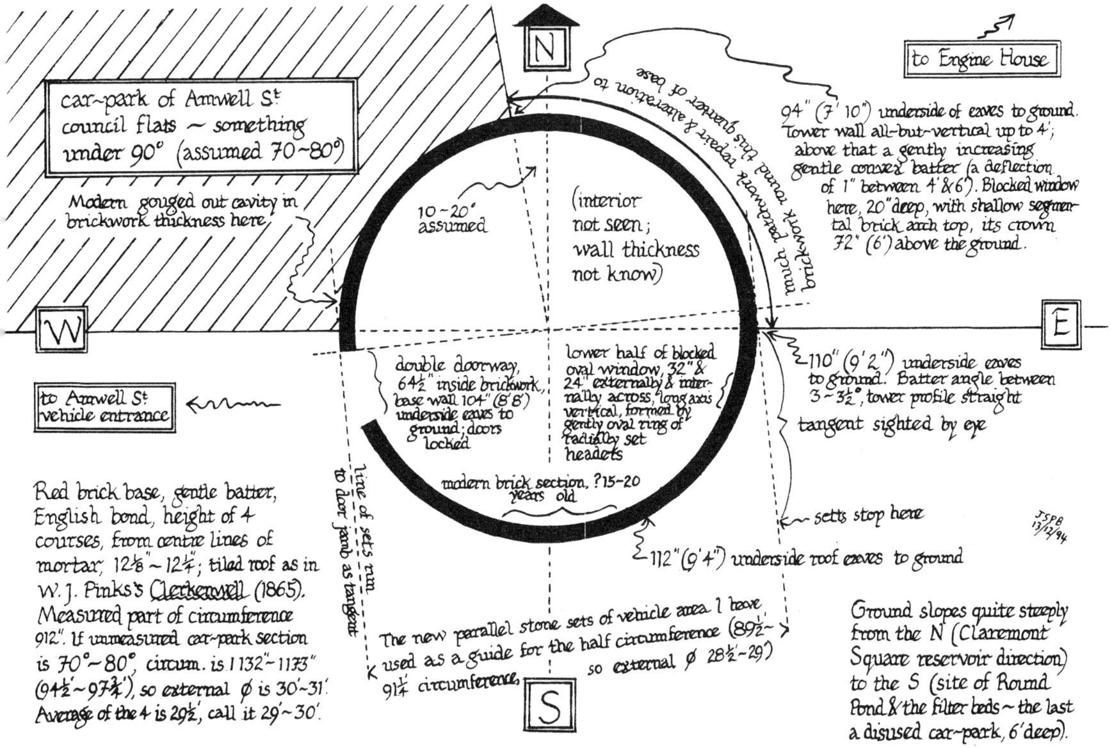

Plan of the New River Head windpump base, from a survey carried out by the author and Mr Stephen Buckland, December 1994. (SB)

The windmills in Finsbury Fields during the 16th and 17th centuries were the best-known in London, and mentioned in two plays of the period. In Chapman's *Sir Giles Goosecap*, first performed in 1606, the lead character tells us, 'I love daylight, and run after it into Finsbury Fields in the evening to see the windmills go', while c1617 Middleton in *A Fair Quarrel*, Act 4 Scene 1, writes 'I have heard 'em roar from the six windmills to Islington'. A windmill is first recorded here in 1556/57; it was shortly joined by another. In 1557-8 the two mills were owned by a former Mayor of London, Sir George Barne, who left them in his will, in which they are called the East mill 'standing at Fyndesburey uppon the dung hill' and the West mill. The former was built on a mound of human bones taken, according to one source, from a local charnel-house, and to another, the churchyard of Old St Paul's, together with human excrement and rubbish collected from the streets. In the same year Thomas Wells was granted a licence to erect a third mill 'upon the Cytyes Ferme Grounde at the Leystowe [laystall: the name given in those days to a rubbish dump] in Fynnesburey Felde on the easte syde of one of Barnes' mills lately erected'. Wells was granted a lease of what was probably the same mill in November 1562.

The reclaimed marshland site was ideal for wind, and the number of mills slowly increased. In late March 1561 four are mentioned in the Repertory of the Court of Aldermen. Shortly after this, one seems to have gone, for a survey in 1567 refers to 'the High Field or meadow ground where the three windmills stand, commonly called Finsbury Field'. Another was blown down in 1599 or 1600 but it was agreed to rebuild it. Attempts to suppress the mills to avoid spoiling the view were unsuccessful, and they continued to flourish; by c1617 they had risen to six.

Thereafter the number is constant; from mid-century it declines. There were six mills in 1658 and 1666, but in 1669 only five. Mr Stephen Buckland states that all the mills had gone probably by 1676, and definitely by 1682, but according to Farries and Mason the last was not pulled down until about 1750, when St Luke's Hospital was built on the site.

The needs of the millers might sometimes conflict with those of local developers, but they usually got their way. In a case in the Court of Common Pleas, a nearby house, which had been built so tall that it cut off the wind from the mills, was ordered to be pulled down.

The technical features, both architectural and mechanical, are analysed by Mr Buckland from the various surviving illustrations:

'We are fortunate in having a good record of the two earliest Finsbury Field windmills, Barne's two, in the copperplate map of Moorgate of c1553-59. Treating them as one, the buck is large — little smaller, I think, though of a different shape, than the latterday post mills of Lincolnshire, Lancashire and Cheshire, or the Midlands. And it is broad, perhaps almost square in plan . . . The ridge of the pent has two finials and a weathervane. At the rear there is a lucam over the square hole for the hoist bollard; the ladder and door are to one side, and the tailpole is held steady by a pair of sheer legs as used today on the mills of French and Belgian Flanders. The ladder has a sack-slide stop, with two sacks visible against it.

'The sheer-legs and stop are very Flemish. The sails are fairly broad, equal double-sided frameworks with hemlaths but no uplongs to the bars. They are typical of English mill sails of the 16th and 17th centuries, but disappeared so long ago that they did not survive to be photographed, though to this day they remain one typical form of French mill-sail frame. They span around 40 feet. The map's draughtsman was thought to be Anthonis van den Wyngaerde of Antwerp, a copperplate engraver and cartographer who was in the London area in 1559; however, this is now doubted. There is nothing Flemish about the form of the mills, or their sails, all of which may be taken as representing with tolerable veracity what was there; but if a Fleming did draw the map, were the sheer-legs, the sack-stop, and, possibly, the hoist lucam added by him from what he knew at home? They were standard Flemish features of the time, or at least shortly afterwards, but are not otherwise known, to my knowledge, in England.

'In particular, though the sack-stop is drawn with some uncertainty — notably, it is not made clear that it springs from the ladder strings — it has two back braces to the tailpole exactly as in the paintings of the much larger Flemish post mills of Jan Breughel I of c1605-10 (though admittedly this is the obvious way to brace it).

'The rear manual hoist is an old feature: it appears in six of the late 13th to mid-14th century manuscripts of post mills viewed from the rear that I have seen, but all are in manuscripts from north-western Europe. However it appears (without a lucam) on the probably 16th century Greenford windmill and is explicitly mentioned on an Essex windmill in a tale in print by 1635, but that is all I know for an English windmill around this time, and indeed for any earlier period.

'Each mill is on a low, small, but very distinct mound. Posts indicated round the mills seem too close in to serve as attachment points for a portable winch, and are probably to keep people and animals away from the sails. The ladder and tailpole extend beyond the mound and posts. The quarterbars on the left-hand mill have a marked concave bow, which is unusual; and I suspect, by comparison with other pictures, that the apparent blocks that the quarterbars die into are in fact the ends of massive ground-laid crosstrees, which have been drawn too big in the right-hand mill.

'The maps of Faithorne (1658) and Leake (1669) confirm that the Finsbury Field mills are post mills to the end, but otherwise they are shown as well drawn but more or less conventional post mill symbols.

'Fortunately, we know something of one of the latter mills. In 1651 an anonymous Lambeth-based pamphleteer contrasted the capability of existing corn mills with the claimed performance of his own near-perpetual-motion mill. In listing the disadvantages of windmills, he wrote:

"Fifthly, the Cog-wheels in most Wind-Mills are (in the diameter) 8, foot or under (I take the example from one of the best Wind-Mills in Moorfields, which I think is not much inferior to any in England) the trundle is at the least two foot, which is 4, to one. Now I have observed in a pretty stiffe gale, that the sailes, and consequently the Cog Wheele hath gone about 8. times in a minute, that is 24 (sic — 32) turnes in a minute for the stones: And for the most part of the time they work, it is not much above this rate, which cannot dispatch above 3. bushels in one houre, but a good Mill of mine may grind a quarter (8 bushels)."

'His belief that the mill he went over was one of the best in England is probably true enough. The figure he gave for the maximum rotation rate of the sails is far too low, doubtless because he badly miscounted the length of a minute: 16-20 rpm would be likely, giving 64-80 rpm for the stones. Otherwise, we can probably accept what he says, though he may have rounded down the output a little. This may be compared with figures of 4-8 bushels ground per hour in a good wind with one pair of stones, as published by Bennett & Elton in 1899 for several post mills of the Midlands and the North-West.

'The brakewheel diameter of 8 feet is much the same as today, as is the gear ratio for a direct drive mill.'

A windmill in East Smithfield was held by Thomas de Moose during the reign of Edward I (1272-1307).

CRANFORD

C17 mill. 102774 (approx.)

In 1604 a windmill was included in the conveyance of the manors to Sir Roger Aston, and it was still standing in 1619 when a part of the manor was conveyed to Lady Berkeley. It was probably demolished in 1625. Its position was on the west bank of the river, close to Cranford Lane, and near a gate leading to the common land.

EALING

Early mill.

A windmill stood on the demesne of Ealing manor in 1318, and was rebuilt in 1363-4. The new mill was destroyed in or before 1409, and had not been repaired ten years

Mill at Edgware, undated. (PB)

later. It must have been made good by 1431 – when it is again recorded as damaged. After this, references cease and no other is recorded, suggesting that the lords of the manor, weary at this catalogue of misfortunes, had abandoned their experiment with wind power.

EDGWARE
Unidentified mill.

The only evidence of a windmill at Edgware is a sketch by one of a family of artists called Havell, currently at the Greater London Record Office. From a newspaper or magazine, and undated, it is entitled simply 'Mill near Edgware'. It is a tower or smock mill and appears to be octagonal or hexagonal. It has a domed cap with a short decorative finial and a rear extension for the winding gear. The mill is connected to the nearby house by outbuildings. If the mill did not stand within the Edgware parish boundary, it probably belonged to Hertfordshire.

EDMONTON
(1) Early mill.
(2) Post mill, survived into C20. 329928

A windmill stood on the manor of Edmonton in 1272 and 1295. In 1345, in the Calendar of Charter Rolls, it was out of action through lacking a millstone, and still so in 1359.

Over two hundred years later, a mill was erected on the Weir Hall Estate between 1605 and 1627. Still standing in 1801, it is shown on the inclosure map as to the south of Silver Street, near the watermill (at approximately 328926). According to the *Victoria County History* it was replaced some time before 1819 by the post mill which then stood on the north side of Silver Street, on the site of the later Windmill Road. Surviving photographs show features characteristic of earlier post mills. It must have been moved from another site. Possibly it is that of 1801, relocated; if so, it was probably rebuilt at some stage.

The miller in 1823-4 was Thomas Hinckley, by 1845 succeeded by George Hart. From 1851 to the late 1870s the mill was operated by the Parfrey family, as 'Parfrey's

Mill'. Edward Parfrey was the miller in 1852, followed by Philip Parfrey (1862), Youngs Parfrey (1867), and R. A. Parfrey (1874). By 1867 a steam mill had been built alongside the windmill, which after this date was probably used less frequently. In 1878 the miller was John William Neave, still there in 1890, when John Stock was helping to run it; he later took over completely.

Stock is last given as miller in 1899 after which references cease. A postcard of 1904 shows it in a derelict state, with great gaps in the body and the sail frames removed. The steam mill also appears to have fallen into disrepair, with a number of boards missing from its lucam.

The mill collapsed or was demolished some time after 1920. At the end little more than the skeleton remained, with all the weatherboarding missing, though the stocks and whips were still attached to the windshaft. The granary survived until pulled down in the late 1950s.

A few reminders of the complex lingered, though not for long. Iris Krass, in an article in the *Middlesex Review* of August 1965 entitled *The Last Wyndmilne of Edmonton*, reported the demoliton of 'a very ramshackle tower-like structure', which she described as the last remains of Edmonton's windmill — possibly one of the chimneys shown in the drawing in this book. The mill site is now a playing field used by the nearby school.

The mill had four double-shuttered patent sails, on which the leading edges were narrower than the trailing edges, unlike those at Hounslow where they were of the same width. It was winded by a fantail mounted on the access ladder to the

Edmonton windmill in working order, c1890, with the mill house, steam mill and chimney, from a photograph in the *Middlesex Quarterly*, 1954. (PB)

superstructure in the East Anglian fashion. The pitch of the roof was almost straight and the weather beam was corbelled out; both these features are associated with 17th and early 18th century post mills, and are another argument in favour of the mill dating from this period. The weatherboarding on the roof slightly overlapped that on the sides of the mill — another East Anglian characteristic, and an example of how windmills near county borders were influenced by millwrighting trends in neighbouring regions. A splendid watercolour now at the Forty Hall Museum, Enfield, shows a weathervane above the rear roof gable and a lever and chain (or rope) arrangement mounted on the tail, which may have functioned as a talthur or been a kind of sack hoist.

The photograph of the derelict mill shows that the body was extended at the tail at some point during its working life; the extension corner posts are breaking away, returning the mill to its original length.

Edmonton was the last windmill to remain in anything like a complete state in Middlesex, and is still remembered by a few elderly residents.

ENFIELD

(1) Early mill.
(2) C17 mill, Windmill Hill.
(3) Post mill which evidently stood on site of (2) during the eighteenth century, and was possibly the same mill.
(4) Smock mill, the successor to (3). 317967

There was a windmill at Enfield in 1284, held by Henry Frowyk, its exact location unknown. In 1635 one on Windmill Hill, west of Enfield Town, on the eastern edge of the forest known as Enfield Chase, was in good repair. This site was occupied by windmills for at least 250 years.

From 1720 the Enfield windmills were worked by the Robinsons, today still a well-known local family, who originated from Scotland. How they came to be associated with the Enfield mill is the stuff of fairy tales. The first of the Enfield Robinsons, one of a large family (the youngest perhaps?) walked all the way from Scotland in search of a living. Several days after setting off, and as the light was fading, he happened to be approaching Enfield mill. The sight of the mill must have been strange to him, coming as he did from a technologically backward land where watermills were almost universal. Perhaps his thoughts were like those of Jeanie Deans in Sir Walter Scott's *Heart of Midlothian*, who carried out a journey to England and wrote home to say, 'I have seen many strange things which I trust to tell you one day . . . all around . . . are mills whilk havena muckle wheels nor mill-dams, but gang by the wind . . . strange to behold'. Robinson had only five shillings in his pocket. He saw a light, knocked on the miller's door to ask if there were any possibility of employment, and was taken on. According to an account of the story in a local newspaper he later married the miller's daughter 'And when the old man died, became master of this little domain perched high up on the ridge overlooking Enfield Chase'.

Another story associated with the Enfield windmill involved the Earl of Chatham, formerly William Pitt the Elder, the great statesman and Prime Minister. Chatham lived for a time at nearby South Hill Lodge, from which the mill could be clearly seen. He felt its colour spoilt his view and paid Robinson a considerable sum to have it repainted. The following morning, however, it appeared the work had not been carried out. Being a post mill, it rotated in order to keep facing into the wind and, when

Chatham had first set eyes on the mill, and been offended by its hue, he had only seen one side of the structure – something Robinson had mischievously neglected to tell him. He now had to spend another large sum to repaint the other side!

Sadly, the authenticity of this story is doubtful. According to the late H. E. S. Simmons it has been told many times about windmills elsewhere.

It is not known whether the mill shown on Faden's map of 1790 is the post mill or the smock which succeeded it. One account says the former was demolished, another that it was blown down. The new mill became an important feature of the locality, and was frequently photographed. Its site can be identified today, by using the nearby church, which often appears in views of the mill, as a reference point. Joseph Robinson was the miller in 1823. He died in January 1849 and in 1851 and 1855 directories gave an I. (Isaac) Robinson as miller. In 1862 it was Benjamin Robinson, another son of Joseph. By 1885 Benjamin decided to give up milling, for parts of the mill's machinery were advertised for sale in *The Miller* of 2 February. These were the three pairs of stones, measuring 3 ft, 3ft 10 inches and 4 inches in diameter, 4 spring sails and the windshaft. He was evidently unsuccessful, for the mill was used by the family for a few more years.

The date it stopped is uncertain. The 1897 Ordnance Survey map marks it as unused, but Isaac Robinson is still listed as miller in 1902, though only in the local directory, suggesting that trade was by then slack and met only local consumption. The mill was demolished in September 1904. Towards the end of its life it was showing clear signs of decay. It stood without sails for a short time prior to demolition.

The wealth of photographic evidence facilitates a good picture of what the mill looked like. The smock tower was broad, though not especially tall. There is no brickwork in the structure apart from the foundations, the woodwork of the smock beginning at ground level. This is a feature of certain smock mills in Kent and also in the Essex/Hertfordshire region, which this part of Middlesex borders. The cap was a horizontally boarded dome, rather flat in appearance, with a short decorative finial and a skirt of vertical boards at its base. Its large rearward extension, which housed the winding gear, was a prominent feature. Unlike the other late surviving Middlesex windmills, Enfield was never converted to fantail winding.

The four single-sided sails, one pair of commons and pair of springs (until shortly before the mill ceased work, when the commons were replaced by springs as the sale notice indicates), were extremely long and would have almost touched the ground. By the end of the mill's working life it was operating on two sails only.

In one or two respects Enfield mill resisted the march of time, compared to those at Hampton, Heston, Edmonton and Kilburn, all of which had fantails and three of which possessed four shuttered sails.

FELTHAM

C18 mill. 094719

Rocque's map of Middlesex, published in 1762, shows a windmill at Lower Feltham.

FINCHLEY

(1) Early mill.
(2) Early mill, possibly same as (1).
(3) Early mill, possibly same as (1).
(4) Near Basings Pond.
(5) Lodge Hill.

In 1252 a windmill was part of the dowry granted to Alice de Basing by her father between 1262 and 1296.
In 1310-/11 Hugh of Arderne and his wife Alice renounced the right to a windmill in Finchley to Simon le Ferour. Another mill, possibly the same, was conveyed with lands in Finchley and Hendon in 1314-15 by Robert Kersebroke to Simon the Marshal and his wife. A mill was also part of an estate conveyed by William de Taillour to Eve of Boltby in 1346-7. Millfields, an area which was part of Bibblesworth Manor from 1365 until the estate was broken up in the late 16th century, indicates that a windmill was situated on the edge of the Common, near Ballards Lane. In 1398 William de Kent had a windmill in the area which he gave to the Rector of Finchley.

Nearly 300 years elapsed before another windmill appeared. In 1627 a piece of waste ground near Basings Pond, west of the road to Whetstone, was granted to Thomas Rawson to build a windmill for the tenants of the Bishop of London, (who owned the manor of Finchley) together with a house for the miller. Rawson, who lived in Hornsey, probably did not operate the mill himself although he was a miller by profession, but was granted a licence to let the mill and land to William Cholmeley for 21 years. In December, Rawson surrended the mill to Richard Turvin. In 1633 the tenant was Thomas Mitchell, Cholmeley having gone. Turvin was followed by Michael Grigg (1635) and Edward Crane (1654). The latter also owned the windmill at Bushey Heath in Hertfordshire. Crane's widow Eleanor had the mill from his death in 1663 until her own in c1676, after which it passed to their son George, who in turn bequeathed it to his sister Ellen or Eleanor Cropper. This lady surrenderd it in 1691. The mill may have been standing in 1734 but appears to have gone 20 years later.

A separate mill briefly stood at Lodge Hill, near the present Bishops Avenue, in what was then a park owned by the Bishop of London. It is shown on Norden's map of 1648, but does not appear again.

FULHAM

Near Fulham Palace Road. Gone by 1794. 768239

A windmill existed here on the lands of the Bishop of London, who resided at Fulham Palace, for nearly four hundred years. The first reference occurs in 1404 in a surrender by John Schamele along with the field in which it stood, 'Milleshot'. (Variations are Millefield, Windmill Hill and Wyndmylleshot). It covered an area bounded by what are now Harbord Street and Cloncurry Street. The mill stood on the rising ground opposite the Lygon Almshouses, in the neighbourhood of the present day Inglethorpe Street. The road which ran beside it, once known as Mill Way or Windmill Worple, is now Fulham Palace Road. Its position on Faden's map of 1790 is about a third of the way between Putney Bridge Road and the top bend of the river, to the left of the road to Hammersmith.

References in the Court Rolls and other 15th century records are abundant. In 1632 the Churchwardens paid the miller's daughter 'in her necessity' the sum of sixpence. At a vestry in August 1727 it was decided to advertise a reward for information concerning the abandonment of a small child, which the miller nursed, in a sandpit near the mill. In 1738 a highwayman was cornered near the mill and shot himself; he was buried at a nearby crossroads, the traditional place of interment for a suicide.

The mill was well used; in the 17th century people in the Munster Lane area had to be restrained from taking a short cut to it across Fulham Fields. The problem was a persistent one and the order of 1608, which levied a fine on the offenders, was ineffective, for it was restated three times, in 1614, 1618 (by which time the fine had risen to 3s 4d) and 1658.

The length of a tenancy varied from seven, fourteen or 21 yers to as much as three 'lives' (those of the lessee, his son and grandson). The rental varied from 3d a year to £3, usually in cash although flour, grain, salt or fish were sometimes substituted. In addition the tenant was required to furnish the Bishop with a couple of capons, or their value (6s 8d), at the latter's option.

Fulham historian C. J. Feret gives an exhaustive list of the mill's tenants. In 1625 the mill was held by Richard Money who paid 33s 4d. On 27 June 1661 it was leased to Edward Butler, a baker. A new lease was granted to Thomas Dickens of the Middle Temple in 1677, and renewed in 1698-9. Dickens was followed in 1713 by the aptly-named John Mills. In 1748 the mill was unnoccupied, but two years later was leased to Simon Jonas and Shole Focken of Crutched Friars. By 1765 they had been succeeded by Oliver Edwards. In 1771 the 'house and mill' were occupied by Messrs Perkins and Spencer, and in 1781 the lessee was John Waters, who insured the mill against fire for £400.

In 1781 another lease, for 21 years, was granted (again) to Oliver Edwards, but the mill did not survive for long; it was dismanted by 1794 due to its decayed condition. According to the convenants of the lease the heir at law should have rebuilt the mill, but this did not happen due to a dispute between the executors. Before its demise its annual value was £25.

The mill lingered in memory for some 100 years in the names of local buildings: a beer shop called The Windmill once stood to the south of the Lodge, and adjoining it were four houses known as Windmill Cottages. They were demolished about 1880. In 1900 there was still a Millshot Farm on or near the site. Excavations carried out in 1973 found no trace.

HACKNEY

(1) Mill at Stoke Newington.
(2) Union Mill, Broad Street. 341834
(3) Horizontal mill in Kingsland Road, near Regents Canal.

At least three windmills existed in Hackney. The first stood on the Pulteney family estate at Stoke Newington, west of Green Lanes. When the estate was sold in 1810 on the death of Sir William Pulteney, details circulated to prospective purchasers included 'the valuable building materials of a large windmill, begun to be erected thereon by the late [Sir William] but not completed'. The purchaser took over construction, finishing it by 1813, but it was in ruins by 1852 and nothing is heard of it afterwards.

Another windmill stood in the former Broad Street, off Hackney Road, and was known as the Union Mill. It is first heard of in 1833 when the partnership between

Horizontal windmill on Mitcham Common, Surrey, which that at Hackney may have resembled, from an undated engraving. (PB)

Henry and John Harvey, described as millers and drug grinders, was dissolved. In 1838 Daniel Dale, who once worked the mill, was listed as an insolvent debtor in the *London Gazette*, which also describes him as baker and millwright. Another who may have operated the mill was Joseph Jakens, miller and baker, who lived at No 29 Hackney Road and, like Dale, ended up insolvent and in debt. By 1846 the business had been expanded and improved by the addition of a steam mill. It was auctioned that year at Garraway's coffee house, described as a 'leasehold windmill . . . with steam and drug mill, two cottages and ground adapted for erecting other buildings held for 60 years, at a low rent'.

The next two partnerships were also dissolved. In February 1851 that between Thomas Rossetter and Samuel Gates of the Union Wind and Steam Mill, as it was by then known, was terminated by mutual consent. By March of the following year Rossette was in a debtor's prison. Samuel Gates then formed a partnership with his brother, also Samuel, and George Baker, described in the *London Gazette* as a 'journeyman miller' (itinerant). The Gates brothers later left, after which Baker ran the mill in partnership with a Richard Bacon and then on his own account before ending up, by November 1851, 'insolvent and in debt'. Yet another Hackney miller to suffer this fate, in 1853, was Joseph Bradstreet, although there is no evidence that he worked at the Union Mills despite his having lived close by at No 20 Great Cambridge Street. This succession of dissolved partnerships and insolvent millers suggests it was rarely profitable. The mill had probably gone by 1856 when it is absent from Crutchley's map.

The third windmill was of a type rarely encountered in the British Isles. It stood near the Regent's Canal, now the Grand Union Canal, in Kingsland Road, in 1828 and was a horizontal mill — that is one where the sails are mounted horizontally on a vertical axis, (although windshafts are rarely truly horizontal but inclined at a slight angle). The first windmills to be recorded anywhere in the world were of the horizontal type. They were in what is now Iran, where some may still be seen at work. The horizontal windmill was widespread in the Middle East, but not so popular in Western Europe where it was confined largely to the Mediterranean countries. Several examples were built in England apart from that at Hackney; at Margate, on Mitcham Common in Surrey, at Hounslow and at Battersea where stood a famous example built by Thomas Fowler in 1788.

The Hackney mill was encased in shuttering and had a series of vanes, mounted on pivots and 'so arranged that they presented a thin edge to the mill on one side of the main shaft and a resisting flat surface on the other'. The cost was comparatively light. R. Thurston Hopkins believed it was a small affair built by a local market gardener for watering his stock (like that at Twickenham). An American technical journal described it as 'A very material improvement on those [windmils of this type] which have hitherto been erected in this country'. K. G. Farries suggested that its success may have inspired the building of the Mitcham Common mill.

HADLEY

(1) C17 mill, Beacon Hill.
(2) C18 mill near the present Drury Road. 247976
(3) Hadley Green. 246976
(4) Hadley Green. Successor to (3). 246976

In 1288 a windmill was the subject of a lawsuit brought by the Abbot and Convent of Walden against Isabella de Frowyk.

In the 17th and 18th centuries there were two windmills at Hadley, both within Enfield Chase and held by Enfield Manor. One was on Beacon Hill and is first recorded in 1629. By 1636, according to a later chronicler, it had for some reason been pulled down and the wreckage removed by one Michael Grigg. At the time it had been tenanted by a man called Hudson. However, it was clearly replaced, for it is recorded again in 1649 when conveyed by Sir Thomas Allen of Finchley to Allen Brent of Friern Barnet, yeoman, along with the half acre it stood on and a house and other buildings. It may have disappeared by 1698 when it is not shown on a survey of the Chase.

The other, which gave its name to Mill Corner and the Windmill Inn, stood a little to the north of the modern Drury Lane near the Great Northern Railway, a few yards within Enfield parish. First recorded in 1635, it disappeared some time between 1740 and 1777. In 1636 it was in the care of Thomas Coningsby, or his tenant. In 1686 William Clarke was in occupation; the following year a question arose as to whether he was liable to pay rates to Hadley on account of the mill. Evidence was adduced that Clarke and his two immediate predecessors, Crane and Reed, had done so, and it was decided that the mill and mill house lay within the bounds of the parish. Clarke, who seems to have been a formidable personality, contested this judgement and somehow managed to get the parish overseer arrested. This did him no good in the end however, for the records show that he continued to contribute to the poor rate, and at a higher level than before. He would seem to have been in the right though, for the survey of

1636 clearly indicates that the windmill was included within the Chase and in a perambulation of Hadley parish in 1772 the boundary line was distinctly drawn through a point facing the spot where the mill had stood.

Some fifty years later, in 1821, another windmill made its appearance at Hadley Green. Almost certainly the last corn windmill to be built in Middlesex, it had its origin in the desire by wealthier householders, who would have supplied the money for the project, to create employment and reduce parish expenditure by providing cheap flour. The mill seems to have proved unsatisfactory, for it was replaced by a new one in 1827. This later came into private ownership. It had disappeared by 1870 when a house was built on the site. John Mallett was the miller in 1852 and F. Green Edwards in 1867.

HAMPSTEAD

(1) Early mill.
(2) Coneyfield, on east side of Frognal. Standing 1597, 1636.
(3) Cloth Hill. 263861
(4) C17 mill, Highgate.

A mill was standing on the demesne of Hampstead Manor in 1270 and was rented at £1 a year. The manorial accounts record a new oak post in 1273. The mill was then worth 24s; this had risen to 26s by 1288. It was blown down in 1294 but rebuilt soon after, the manorial accounts for the following year including '1 standard of schepeschides, 2 crosschidews', and '1 rod carried from Prosendown'.

In 1312 it was valued at £1 13s 4d, and in 1347, 32s. Repairs costing 5s 7½d were carried out in 1347. These included '16 ells of canvas bought for the mill sails 4s 8d; iron mattock and 2 iron dowlegge worked for the cogwheel and for putting on same 8d. 1 lock for the mill door with key worked 3½d'.

The mill was profitable, repaying the cost of repairs. However, like that at Shepperton it suffered from the effects of the Black Death. It was out of use in 1353, the tenant miller having died from the plague; a successor was found, but he fled the area in 1376. John Cogel took up possession at Christmas. In 1377-8 it was worth 22s 6d, the reduction in rent being due to its standing idle 'from the feast of Michaelmas to the feast of Christmas' for want of a tenant. Further repairs were carried out which included: '4 iron bindings bought for binding the spokes of the mill, with nails for the same 3s 4d. 2 new bills [for trimming millstones] bought for the mill 20d. Total cost 5s.'.

Cogel held the mill until 1390 when he was succeeded by John Drew, It was then worth 30s. In 1391-2 repairs amounting to 6s 8d were carried out: 'cost of 1 carpenter cutting down and setting up timber and putting thereof 2 new spikes on the axle of the mill, in all 5s. 8 bindings with nails bought for the same binding the said spokes 20d.'.

At this point Drew defaulted, and there were problems in getting a tenant; none is recorded from 1392 to 1399, when the mill was empty and unrepaired. Another mill stood near the churchyard in 1597; it was in 1636 part of a copyhold estate known as Coneyfield, on the east side of Frognal. In 1661 this mill passed from Charles Purrett to his sister, Bridget Hattan, on the former's death. Apart from the mill there was a garden, orchard and cottage. By c1680 the mill and its associated buildings had been replaced by the Old Mansion, No 94 Frognal. The Frognal mill was one of the two windmills depicted in various views, including those by Hollar and Visscher, between c1593 and 1638. The other, western, mill stood by 1614 on the Heath at the upper end of Cloth Hill. In 1665 the miller was a man named Keyes. In 1674 the mill was owned by Philip Cater and tenanted by Richard Thornton, to whom Cater conveyed it that year. Thornton in turned conveyed it to Richard Snelling, in 1686. Later tenants were George Love (1704) and William Knight (1725). When Knight sold the site in 1728 the mill was described as 'broken', and by 1759 it had gone. The site later became known as Windmill Hill and the house as Windmill House, later Mount Vernon House.

A relic of the Mount Vernon mill was uncovered in 1960 when contractors, building a new house in Holly Place, Old Hampstead, uncovered a brick-lined hole which was probably part of a threshing floor. After the chaff and stalks of corn were removed the grain was shovelled into the hole, which served as a hopper, having converging sides which directed it into an underground passage leading to the mill. Here it was caught in a wheeled receptacle in which it was conveyed along the passage to the mill.

Set into a flight of steps at a house in Heath Street is a millstone, which may or may not have come from the New Grove House mill.

The two windmills at Hampstead would have been a conspicuous sight, and attracted many visitors to the neighbourhood. They were a popular subject with contemporary artists, such as Huych Allaerdt, who depicted them as two post mills in a view of London in 1686. Some controversy surrounds the mill shown in a painting by W. Tomkins, entitled 'High Ground in North-West London', which features a tower or smock mill. The view is said to be of Hampstead Heath, yet by Tomkins' time the Hampstead mills had long disappeared. If the mill was intended to represent an existing one, it is most likely to have been Kilburn.

The mill which appears in an 1830s painting of the Heath by Constable, again of tower or smock type, may be dismissed as a figment of the artist's imagination.

A windmill was built on the north side of Chapel Field, Highgate, shortly before 1601. It had a relatively short life, disappearing by 1641.

HAMPTON

Windmill Road. 139711

Hampton Windmill is an enigma. Firstly, in its appearance and construction it had several striking features which distinguished it from the English norm; secondly, little is known about its history.

In 1785 John Naylor, a millwright from Twickenham, applied at a Manor Court for permission to build a windmill to grind corn, particularly pearl barley and oatmeal, for mainly local consumption. When the mill was partly completed, Naylor ran into financial difficulties and had to abandon the project, which lapsed until a mortgagee took over in 1781.

For a number of years the mill was owned by the Deeks family, who operated it in conjunction with a nearby watermill. The last directory to mention it is *Kelly's* of 1862, which give Zachariah Warren as miller. After this it ceased work. Another miller, who at different times worked both the windmill and the watermill, was James Humphreys, who was there in 1852 but is listed in the *London Gazette* of 19 November as an insolvent debtor. He was succeeded by C. H. Brazier. The property was sold in 1874 and the mill shortly afterwards pulled down; the *Surrey Comet* reported the sad death of a workman who fell from the top while working on the demolition.

Several illustrations have survived, and these show it to have been distinctive. The one photograph, which shows the mill getting into a bad state of repair and probably taken between its ceasing work and being demolished, has been heavily retouched.

It depicts a smock mill which is broad with a steep batter, having a rather square appearance. The corners of the smock, where the edges of the weatherboards meet, are proofed with vertical strips of an unidentified material — probably either wood or lead to keep out rainwater. The lower part of the smock tower is obscured by vegetation, so we cannot see whether there is a stage or any brickwork in the structure above ground level. The cap is of a curious kind, something between the Kentish and the Norfolk varieties in appearance, and reminiscent of those on mills in North-Western Europe. There are four shuttered sails, springs or patents, on one of which a leading board is visible. The remains of a fantail hang down from a platform at the rear of the cap. The wheel beneath the platform, at right angles to the mill body, is presumably part of the winding gear.

The photograph is similar, bar one or two details, to an engraving of a type which became common after 1870. This was evidently contemporary with the photograph, although it shows no sign of a fantail. The cap is winded by what appears to be a Dutch-type braced tailpole, a feature absent in the photograph, and a stage is also visible.

The three lithographs, dated 1823, by the Dutch artist Simon de Koster are probably unreliable. With its concave sides the mill bears more resemblance to mills in Holland than in England. Even though it did exhibit similarities to mills in that country, it is probable that the artist, who did not settle in England until 1788, was influenced by his Dutch perception of what a mill should look like.

HANWELL

Early mill.

A windmill was built at Hanwell by Richard de Crokesley, Abbot of Westminster, between 1246 and 1258. The site was probably the field still known in the early 19th century as Mill Hill Field, on Cuckoo Hill (now Greenford Avenue) just north of Blood Croft, which would have been a good place for wind. The mill was removed early in the 14th century by one of Crokesley's successors, but it left its mark on local affairs. Between 1339 and 1343 it was indirectly the subject of a legal case in which the then Abbot appeared before the King at Westminster, charged with neglecting to keep in good repair a bridge which Crokesley had built to enable those living over the river to get to the mill. The Abbot denied responsibility for its upkeep. The King ordered him to carry out the necessary works.

HARLINGTON

Probably an early mill.

The names Windmill Hill and Windmill Hill Close, as two roads here to the west of Dawley Road, near Pinkwell Lane, were previously known, are the only evidence that a windmill ever stood in Harlington.

HARROW

(1) Early mill.
(2) C17 mill, Sudbury Hill.
(3) C17 mill, Sudbury Hill.
(4) C17 mill, Wembley Hill.

A windmill, the first definite example to the recorded in Middlesex, and a watermill were standing at Harrow in 1236. The former probably stood near the house known as Flambards; in 1353 it or a successor was mentioned in a fine between Edward Flambard and the Vicar of Harrow.

Milling on the Brent was a hazardous business due to frequent flooding. This may explain why the lords of the manor supplemented the watermill with another source of power, and indeed why the windmill came to supersede the watermill by 1348.

There were windmills at Harrow in the 17th century according to three Parish Register entries: the baptism took place on 27 July 1622 of John Hoy of the 'Wyndmylle'; that of 'John of the Millhouse' in Sudbury, and on 31 May 1629, Thomas the son of John Hoy, 'miller' of Sudbury. Two maps of 1695 show two windmills on either side of the road on Sudbury Hill, near the Hermitage, while a third stood at Wembley Hill in 1673, according to another map.

In April 1672 Sir James Rushout, then Lord of the Manor, leased a windmill and Millhouse in Harrow itself for seven years to Bartholomew Brixwell, miller. A letter dated 10 May 1691 to Rushout from his bailiff survives in the Greater London Record Office. In December 1693 a windmill and tenement were leased to Rushout to David Slye, miller, of Sudbury.

HAYES

C17 mill. 095818 (approx.)

Ogilby's *Britannia* of 1675, along with some of his other road maps such as 'The Road From London to Aberistwith', shows a windmill on the north side of the Uxbridge road between the present Church Road and Hayes End.

HENDON

(1) Early mill.
(2) C17 mill, Mill Hill. 223934
(3) Early mill.
(4) Possible mill, undated.

A windmill is mentioned in the Black Survey of 1321 as worth 12 shillings per year, possibly that which gave its name to Mill Hill (originally 'Melnehel'), later the site of the school. A mill is shown there on Seller's map of 1679, and in 1685 one is recorded as held by Robert Crane. It is not mentioned in Messender's Field Book of 1754, which gives a detailed description of the manorial property, so had disappeared. Its existence is commemmorated by Millfield, now a recreation ground, although this does not mark the actual site of the mill, which stood on the high ground at the north end of the Ridgeway, recorded variously as 'Miller's Land' and in the 1754 Survey map as Mill Fields. This lies between Holcombe Hill and the present Hammers Lane.

According to the *Victoria County History* Goldherd's Mill stood in the 15th century between Cowhouse and Glitterhouse Farms, and the name Millfield south of the Bald Faced Stag public house on Edgware Road suggests the existence of a third.

HILLINGDON

Early mill.

A windmill belonged to Colham manor in 1328.

HOLBORN

(1) 313819 (approx.)
(2) On riverbank near Old St Paul's. Standing 1616.
(3) 17th century mill near Old St Paul's.

An indenture of sale in 1507 included the 'sight' (site) of a windmill in St Andrew's, Holborn. A document of 1505 places it in the Manor of Portpoole in the Gray's Inn district, between Gray's Inn Road and Leather Lane.

John Visscher's Panorama of 1616 shows a post mill on the riverbank about halfway between Old St Paul's and Suffolk House. This is evidently in working order, as it is facing the wind judging from the sails of the boats on the river.

A number of old engravings from before the Great Fire feature a post mill on a high structure to the south of Old St Paul's. A square-shaped tower on a print of the Great Fire is described as a 'Waterhouse', suggesting the mill pumped water.

HOUNSLOW

(1) Early mill.
(2) Brazil Mill. Site unknown. Was standing in mid-C18. 115744 (approx.)
(3) Horizontal mill at Syon Park. Site unknown.
(4) Hounslow Heath. Standing C18.
(5) Hounslow Heath, Heston Mill. 119752

In the annals of milling Hounslow has a certain claim to fame. In Elizabethan times its cornfields, now long swallowed up by urban sprawl, supplied the grain whch fed the Royal household.

A windmill was recorded here in 1362.

A curious smock mill is depicted in an illustration of 1757 which bears the title 'Brazil Mill, Hounslow Heath'. This suggests it was used to extract dye from Brazil wood. It stands on a square building of about three storeys, whose roof forms a stage. Other, smaller buildings are built onto this. The cap is conical, rather like that of Shipley mill in Sussex, and winded by a tailpole, and the alignment of three visisble sails suggests there were six. Smock mills were still a comparative rarity in the mid-18th century, as were mills with more than four sails.

Rex Wailes mentioned, in a paper to the Newcomen Society in the 1950s that, a horizontal mill stood on the Duke of Northumberland's estates at Syon House.

Hounslow had a tendency to produce unusual windmills but Heston Mill (standing some distance from that place) was relatively conventional.

It was built in 1818 on the Heath to the north of the Staines Road, probably as a replacement for one shown at the site on the 1816 Ordnance Survey map; a painting by Richard Wilson in 1765 features an open-trestle post mill.

The mill was put up for sale in 1829, when occupied by John Bryant. The *Sussex Advertiser* states it 'has been built about 10 years'. It had patent sails and a fantail and worked three pairs of stones; the latter would have been arranged two side-by-side in the breast and one in the tail, with those in the breast driven *via* a wallower and great spur wheel and those in the tail from a tail wheel mounted on the windshaft rearwards of the brake wheel. Attention was drawn to the spaciousness of the premises: '[There is] sufficient room in the mill to contain about 150 quarters of corn and do the business pleasantly, and a warehouse capable of holding the same amount'. Bryant failed to get a purchaser for the mill and was still there in 1834, but he was succeeded by Joseph

Cooper who ended up in prison for debt. By 1854 the mill had passed to the Marshall family; initially William Marshall and his brother Thomas ran it jointly, but their partnership was dissolved in May 1854, William afterwards operating on his own until his death in March 1862. It was then taken over by William's son James. The Marshalls were running the mills in 1890, but soon after that they passed to a family named Ashby. By 1891 a large steam flour mill had been built close to the windmill, and progressively took over.

On 24 August 1895 the mill and the other buildings on the site were destroyed by fire. The following account appeared in *The Miller* of 2 September.

'BURNING OF HESTON MILLS, HOUNSLOW. In the early morning of Saturday August 24th, a disastrous fire broke out in Mr Ashby's Heston mills, at Hounslow. These mills consisted of a row of buildings forming a sort of parallellogram, with outbuildings standing apart from the main row. At one end was the warehouse, then come the mill proper, which contained a 5-sack roller plant, and then a row of dwelling houses and cottage. In front of the mill and dwelling houses stood a windmill, as well as 2 or 3 buildings used as a shop, store and office. The mill had been working until 6 pm on Friday 23rd, when the staff, with the exception of one man, left off work. A gang of millwrights had been busy in the mills for some days previous to the catastrophe, and on the Friday they had been putting the finishing touches to their work, but they also left the building at 6 0'clock this afternoon. From 6 till 10 pm the premises were in the charge of the engine driver, who was looking after some millstones which were at work on barley. At 10 o'clock all seemed to be safe in the mill, and the engineman shut down and went home. But about five minutes past one on the following Saturday morning a passer-by noticed flames shooting from the warehouse lukum, and promptly aroused Mr Ashby, who lived in the dwelling house by the mill. A party of soldiers turned out of the neighbouring barracks and tried to extinguish the flames, which seem to have spread very rapidly. Before long as many as six steam fire engines were on the spot, but there was a delay in getting to work on account of the scarcity of water. The well which supplied the mill was not accessible on account of the smoke, while none of the brigades present appear to have been provided with any means of raising water by suction. The town main was some distance away, and the hose had to be carried over several fields before a connection could be effected. Altogether the fire seems to have had a good half-hour's start, which enabled it to take a firm hold of the mill as well as of the granary in which it apparently originated. The flames spread into the mill in the teeth of the wind, which was against them, and also across the yard, catching the windmill and the group of buildings used as offices and stores. These structures were all totally destroyed, only the dweling house and the cottage were saved from the fire, but damaged by water. No cause was assigned for the fire, which seems to have started in a room that only contained an oat-crusher, a wheat-grading sieve and a sack-hoist. The damage will reach £13,000.'

According to another account, in the *Middlesex Chronicle*, firemen from Isleworth, Brentford, Richmond, Chiswick and Teddington were involved, and people had to be evacuated from nearby properties with the help of the troops. The fire was so intense that the ruins smouldered for a week. A leader article later appeared in the *Chronicle* castigating the fire brigade for their incompetence.

A photograph in the Treaty Centre Library at Hounslow shows the aftermath, with the ruined walls of the roundhouse, along with the gutted and partially collapsed steam mill building.

The large body of the mill stood above a two-storey roundhouse whose roof was painted white, and which had a loading door in the upper story. The weatherboarding on the lower part of the breast was carried down to form a shield to keep rainwater from entering the gap between the body and the top of the roundhouse roof, and was stepped back where necessary in order to clear the latter. The four double-shuttered patent sails were of the East Anglian type, with large reinforcing clamps and trailing and leading edges of equal width. The body was turned into the wind by a fantail mounted on an unusually tall carriage, whose height may have been intended to counter the considerable weight created at the front of the mill by the four large patent sails and the two pairs of stone in the breast. It is not clear whether this was attached to the ladder in the Suffolk fashion, or the tailpole, which was here retained, as in Sussex, although it is clearly braced to the latter. A number of battens were fixed to one of the upright members of the carriage to provide a simple ladder giving access to the fantail for maintenance.

This mill was not typical of the county. A millwright from East Anglia evidently had a hand in fitting, or at some point replacing, the sails, while the arranging of the weatherboarding on the lower breast to form a barrier against the rain was reminiscent of certain post mills in Sussex and Kent. The patent sails, fan and three pairs of stones reflected an advanced millwrighting tradition.

ISLEWORTH

Early mill.

In 1370 there was a windmill on the manor of Isleworth, but it may have stood on the Twickenham portion.

ISLINGTON

(1) Early mill, Canonbury.
(2) White Lead mills, Rosemary Branch. 328839
(3) Windpumps at Islington cattle market.
(4) City Road

In 1306 the manor of Canonbury had a windmill worth 40s a year.

In 1786 Samuel Walker, an ironmaster from Masborough, near Rotherham in Yorkshire, built a windmill at the White Lead factory at Rosemary Branch in Islington. A second mill followed in 1792. They were tower mills, with domed caps winded by fantails whose platforms were long and positoned horizontally. They reflected the millwrighting practices of Walker's native county and were more advanced technologically than most windmills in southern England. They attracted much local comment. Fantails were still a rarity, as were the five sails of one of the mills; four sails was generally the norm. They were also extremely powerful; a plan shows one had eight pairs of stones.

By 1811 the five-sailed mill had been converted to a four-sailer. A five-sailed windmill had to cease work if one was damaged, whereas a four-sailed mill could carry on working with two.

They were still working by wind in 1822, for an etching made by Jane Smith shows them with sails and fantails in place, but by 1835 these had been superseded by a 20 horsepower steam engine. A contemporary watercolour by C. H. Matthews has them without sails or fantails; the five-sailed mill has also lost its cap. The four-sailed mill retains the iron casting or cross by which the sails were attached to the windshaft

— a feature almost universal among windmills on the North-East. The mill towers were integrated structurally with the other works buildings.

The mills were both standing in 1868, when they are indicated on a plan of the works by two circles. At some time one was pulled down and the other reduced in height. The latter's truncated tower survived until 1945 when damage by a V2 rocket necessitated reconstruction of the works, and the remains were demolished.

In complete contrast to these huge tower mills were the two small windpumps erected by 1820 at Islington cattle market. Their purpose was to supply water to the covered enclosures or 'lairs' which housed the 4,000 or so cattle. They took the form of hollow-post mills, mounted on low square bases which were nicely ornamented, and had four sails. Each was situated over a well from which the water was pumped to the cattle troughs *via* pipes. The *Mechanics Magazine* of 27 May 1837 describes their workings:

'In the centre of a stone building of a quadrangular form, is erected the pump (a) of cast-iron, secured by strong iron braces, and carrying on a flange at its top, an iron frame (b), which supports the crankshaft of the pump (c), and also the vane-axle (d); ee are the vanes. A pinion (f), on the vane-axle, takes into and drives a spur wheel on the crankshaft of the pump, the piston being attached to the crank by slings, and the piston-rod working in a guide, which preserves its parallelism; (g) is a brakewheel for impeding, or altogether stopping, the motion of the machine. The revolution of a set of vanes (h) causes the whole of the mechanism to rotate on the top of the pump and continually occasions the large vanes ee to face the wind. A good well supplies each pump with water, which is from thence distributed by pipes into troughs in each of the cattle yards throughout the market'.

A windmill was built in or shortly before 1890 on the roof of a warehouse in the City Road for Messrs Carwardine & Co, then a well-known milling firm, to operate machinery for the production of wholemeal flour. The millwrights were Messrs A. Williamson & Co. It was of the type patented by an American named Halliday and frequently used for pumping water, with an annular sail mounted on an open wooden framework which stood some 50 ft above the warehouse. *The Miller* of 6 October 1890 reports that the mill could give out 8 horsepower when the wind was travelling at 18 miles an hour. Possibly its demise was connected with action taken by the council in 1892 because the mill carried advertisements. It would certainly have been ideal, for it must have been conspicuous.

A history of Islington also states that a windmill once existed in the parish of Newington. Its position was possibly on the west side of Green Lane, near Newington Green.

KENSINGTON

C17 mill, Hyde Park.

In the 1660s a post mill appears as a distant feature in a pen and ink drawing of Hyde Park by the Dutch artist Michiel van Overbeek, standing outside the boundary wall of the Park on the ridge which then formed part of the ancient manor of Knotting Barns, now Notting Hill. It was presumably one of the 'mills' mentioned in a document of 1543 as belonging to the Manor.

In 1651 ten acres of arable land called Windmill Hill were among properties sold by Sir Arthur Gorges to Sir Michael Warton.

KINGSBURY

(1) Redhill. Gone by 1754.
(2) Possible mill on site of Willesden Cemetery. 218845 (approx.)

A windmill stood near the Edgware Road at Redhill around 1675. It was there in 1684 and between 1729 and 1738, along with a cottage, but had been blown down by 1754. Another mill may have stood on the site of the present Willesden Cemetery, which was previously known as Windmill Hill.

LALEHAM

Early mill.
(2) Standing 1816. 705055

There was a windmill at Laleham in the 14th century, and a mill of unspecified type is mentioned in a grant of the site of the manor in 1608. There was at one time a Windmill Hill. Later a windmill is marked on the 1816 Ordnance Survey map, and in 1832 and 1834, directories mention Thomas Ryde of Laleham, miller. The mill stood at the end of a short track leading from the west side of Ashford Road, about halfway towards Staines Road.

LIMEHOUSE

(1) Sawmill, standing 1663.
(2) Sawmill, standing 1768.
(3) Smock mill, standing 1791.

In 1663 some Dutch millwrights built a windmill at Limehouse to drive circular saws. It aroused the opposition of local sawyers, who feared that this mechanisation would put them out of business. They damaged the mill so badly that it had to be abandoned. History was to repeat itself a hundred years later when another wind-driven sawmill was erected by Charles Dingley, who employed James Stansfield, an authority on the sawmills of Holland, on the project. Shortly afterwards, in May 1768, the following advertisement appeared in the *London Gazette*:

'Whereas it has been humbly represented to the King, that a wind saw mill at Limehouse, belonging to Charles Dingley Esq., was on Tuesday the 10th inst. unlawfully and with force begun to be pulled down and demolished, and was thereby rendered totally useless, by a mob of people unlawfully, riotously and tumultously assembled together, to the disturbance of public peace; His Majesty, for the better discovering and bringing to justice the persons concerned, is hereby pleased to promise his most gracious pardon, and also a reward of £200, to any of them who shall discover his Accomplice or Accomplices therein, so that he or they may be apprehended and conviced thereof . . .'

The mill of 1663, and perhaps also Dingley's, considering Stansfield's familiarity with Dutch sawmills, probably resembled the 'paltrok' sawmills then being built in Holland, as did that which stood at Lambeth in the early 18th century and appears in an engraving by Schenk. This type of mill had a distinctive appearance unlike that of most other windmills in either England or Holland. It resembled a four-sided smock mill and had a curiously-shaped cap which was integral with the main structure, since the whole of the latter, which rested on a low circular base of brick or stone, could be turned into the wind, whereas in most smock mills only the cap is rotatable.

According to R. Thurston Hopkins in *Old Watermills and Windmills* (1929) the mill was later rebuilt, and this time went unmolested. A smock mill is shown rather indistinctly on a 1791 watercolour by Chieveley and, unless this was one of the Millwall mills, it could have been the reconstructed sawmill.

MARYLEBONE

Standing 1708.

A windmill here in 1708 appears evident from a curious symbol which occurs on a map of the parish in Marylebone Library — a house above whose roof protrude two windmill sails in the form of a V, and which seems to stand on the trestle of a post mill. The land on which the mill stood was then part of an estate owned by the Duke of Portland.

LEFT: A Dutch 'paltrok' sawmill. (PB) RIGHT: Detail from a 1708 map of the Duke of Portland's estate at Marylebone. (PB)

SOUTH MIMMS

(1) Early mill.
(2) Early mill.
(3) Unidentified mill, standing 1628.

A windmill may have stood in the demesne of South Mimms by about 1220, when Gervase and Arnold, millers, were recorded. It would have been among the first windmills to appear in Middlesex. It was recorded in 1336 and again in 1349, when it was worth 13s 4d. Its site may have been in Windmill Field, somewhere near the manor house.

The mill owned by Isabel Frowyk in 1289 probably stood somewhere between Old Ford and Christ Church. It was among the appurtenances of Old Ford held by the Frowyks in 1310. In 1639 the same mill or a successor was leased to Thomas Allen. Sometime after, it disappeared.

In 1628 an unidentified mill was the property of Steven Bowman; in 1668 it was granted to William Bowman, presumably a descendant of Steven, by the Earl of Salisbury, a prominent local landowner.

MONTHILL

17th century mill.

According to Bennett and Elton it was ordered during the reign of Charles I that a windmill which had been moved from Monthill should be brought back.

NORTHOLT

(1) Dabb's Hill Lane. 1386 8552
(2) Mill Post Field. 1250 8250

Northolt has possessed two windmills. One stood near the present Park Station, in Dabb's Hill Lane (a corruption of Dabb's Mill Lane). Another was in Mill Post Field; almost certainly a post mill; the substructure perhaps remained for a time after the rest had gone.

NORTON FOLGATE

A smock mill stood here at one time, according to H. E. S. Simmons. The late T. J. Mason possessed an illustration, which has unfortunately disappeared along with the rest of his extensive collection of material on the area's windmills.

PERIVALE

(1) Early mill.
(2) Standing C18 according to folklore. Possibly a successor to (1).

The existence of a windmill on the manor of Perivale in 1342 is regarded by local historians as evidence of contemporary prosperity. There was also a mill there later. According to J. Allen Brown in *The Chronicles of Greenford Parva,* or *Perivale, Past and Present,* (1890), a woman in her eighties remembered being told in her childhood by old people that a windmill stood in the area near an ancient oak tree, and this survived until the early 18th century.

Windmills have often features in tales of a romantic or supernatural kind. Allen Brown's book includes one.

Some 250 years before he was writing, a windmill stood at Perivale. The miller was an unsociable man named Abel Reed, who lived alone. Nearby stood the cottage of Dame Gigs, an old widow regarded as a witch, and therefore shunned, by the villagers. Reed believed her to be a charlatan and took every opportunity to tell her so. One day he happened to meet her while coming home from a drinking session. He was in a bold and argumentative mood from the effects of too much alcohol, and began behaving threateningly. She was so angered that she cursed him, declaring that within the next twelve months he would be either drowned in the Brent or crushed by his own millstones. From then on the villagers regarded him as tainted and would have nothing to do with him, so that his business fell off.

About a year after this incident Reed disappeared, and a search was mounted. Mindful of Dame Gigs' prophecy, the villagers were reluctant to enter the mill to seek him out, with the exception of the old woman herself. Climbing a tall ladder inside, she slipped, fell and was badly hurt. She was taken home while the others ventured fearfully in search of the missing miller. Between the millstones were found pieces of rag and broadcloth, while pulverised bones, blood and hair told too plainly what had become of the unfortunate Reed. Dame Gigs died shortly afterwards.

After these events no-one took on the mill and it became a ruin. No villager would pass the spot at night unless they had to, and those who did later told their friends how they had heard the moanings and shriekings of the ghosts of Abel Reed and Dame Gigs, and seen them pursue each other around the gallery of the mill. Eventually an old miser declared his intention of occuping the mill, and despite warnings did so. Shortly afterwards he went missing too. A few villagers went to the mill and called to him. There was no response. A young man named Simon Coston, an orphan with no friends or relatives, entered the mill, and a few moments later Coston ran out exclaiming 'The ghost! The ghost! Fly, fly for your lives!' The villagers ran off.

Soon afterwards Coston in his turn vanished. A group of men, fortified by drink and armed with guns and pitchforks, returned and shouted the names of the missing men. They were answered by a hollow, sepulchral voice which bade them

'Bury me, bury me, ere you sleep,
In Perivale churchard six feet deep.'

Inside they found the dead body of the old man; of Simon Coston there was no sign.

Some 30 years later, a gentleman named Simon Coston purchased an estate at nearby Greenford and built himself a fine house there, where he lived with his wife and family. A few of the older villagers detected in him a certain resemblance to the poor orphan, but for the most part no-one suspected that he and the wealthy Squire Coston were one and the same.

Coston died shortly after 1665, apparently by committing suicide, and his executors found that he and the orphan were indeed the same, and they also discovered the key to the mystery of the mill. When Simon entered the mill he saw that the old miser had died while counting his money. He hit on the plan of frightening the others away and then securing the money. Having transported the miser's body to where it was found, he concealed himself and, when the neighbours came, uttered the doggerel, and decamped. He went to London, set sail for Flanders, gained work in a merchant's house, became a successful merchant himself through hard work and business acumen. Having made a handsome fortune he determined to return and enjoy it in his native place.

Although at one point so rich that his horses were said to be shod with silver, Coston seems to have run into financial difficulties. He also lost his wife in the plague, among other misfortunes, and eventually drowned himself in a pond in the grounds of his house.

According to Brown, Coston's house was destroyed by fire, but the fishpond remained, as did the avenue leading to the house, known as Coston's Lane. Many inhabitants had a strong aversion to going by the Lane at night, for fear of meeting Coston's ghost, although an old waggoner said that he had done so many times and 'never seen anything worse than hisself'. One man declared that on going to the field where the pond was, his horses trembled in terror, and he saw 'something white' hovering over the pond, making a great splash as it jumped into the water, just as Coston did over 200 years before.

PINNER

(1) Early mill.
(2) Standing C17 and after; latterly replaced by a smock mill. 112903

In c1285 a windmill on the Manor of Harrow was worth five marks a year. It stood on the lord of the manor's estate at Woodhall, to the north of the Uxbridge Road.

On 7 May 1619 the Lord of the Manor granted William Crane of Stanmore an area of waste ground on Pinner Common for 70 years at a rental of 20 shillings a year on condition that he built on it within 12 months 'One good stronge and substancialle wyndmylle'. Crane worked the mill himself and his descendants carried on; in 1662 a house built beside the mill was occupied by Thomas Crane, and the family were well-known in the milling trade, owning and operating several windmills and watermills in the area.

By 1692 the mill was in need of repair, and the Lord's bailiff wrote to him on 14 May informing him that a new millstone and 'axle' (the latter presumably meaning the windshaft) were required, the millstone costing £10. The windshaft required only the trunk of a single tree 'as David [the miller] had lately', and this could be provided free by the Steward of the manor. Before the work could be carried out the miller died and the bailiff engaged a carpenter named John Pond to do it. It proved worse than envisaged, and the bailiff feared that, if not rehabilitated, the mill would collapse or be pulled down by poor inhabitants for firewood.

On 10 May following Pond took out a 41-year lease at a rent of £3 per year. He had to keep in good repair 'all manner of iron worke, timber sayles, saylecloths, wheeles, milstones and all other things and utensills'.

A strong wind blew down the mill early in 1721 and the site was subsequently granted to John Blake, another carpenter, for £36 15s, on condition that he built a new windmill. When Blake died in 1743 his elder son, John, took over the carpenter's business while a younger, Samuel, became responsible for milling. The next three owners were William Jessup, who bought the mill from Samuel Blake c1752, Nehemiah Nesbit, to whom Jessup sold it two years later, and Thomas Nesbitt, Nehemiah's nephew and heir, who sold his inheritance to Joseph Dell in 1775. In view of their occupations — Jessup was a soap-boiler, Nehemiah Nesbit a merchant, and Thomas Nesbit a turner — they probably employed someone else as miller.

Carpenters returned with Dell, who died in 1784, and his sons (or grandsons) William and Daniel, who eventually took over. It was probably during the Dells' ownership that the mill was rebuilt as a smock.

Thomas Catlin of Watford acquired the site in 1826 and engaged as millers the Jaques family, from Kent. Charles Jaques was miller then, followed in 1841 by James Jaques, assisted by his sons, Charles and James Jnr. By 1851 he was succeeded by James Jnr, who ran the business on his own. James was followed by his son, William.

Shortly after buying the mill from his landlord in 1867, William installed steam-powered machinery and probably an accident with the boiler caused the mill to burn down one night in 1872. William then gave up milling, and the site became a farm, cleared away in the 1960s to make way for a council estate. One of the roads is named Mill Farm Close, while part of the old millstone is set into the base of a slide in a children's playground.

From a watercolour of 1872 the mill stood on a fairly substantial brick base without a stage. The cap appears roughtly similar to the Kentish type. The mill was equipped with a fantail, and the painting shows what appear to be the frames of four patent or spring sails. There is a sketch by B. C. Dexter from the front which also suggests a Kentish-type cap, along with four double-shuttered sails of the Kentish variety, with narrow leading edges. Dexter was active in the 1890s, some twenty years after the mill's demise.

POPLAR

(1) Millwall. Ten, possibly more, mills, standing between approximately 370805 and 375785.
(2) Kerby Street. 378814
(3) Blackwall. Standing 19th century.

Many must have often wondered at the origin of the name Millwall, given to the district around the west side of the peninsula known as the Isle of Dogs. Here, on the river bank or 'wall', there stood what were probably London's most famous windmills. They had their origin in the expansion of commerce after the Restoration, following the upheavals of the Civil War and Commonwealth. The low-lying underdeveloped and extremely fertile marshland was an ideal place for windmills. All were built near the river to take advantage of the obvious opportunities for waterborne trade.

Post mill at Millwall, from a painting by Miss H. J. J. West, 1830. (PB)

The area remained undeveloped until comparatively late, and the Millwall mills were the last in the capital to cease work.

At least seven were built; some old prints show more. The maximum standing at any one time was probably 10. Nine appear on 'A Prospect of Greenwich & London taken from Flamstead Hill in Greenwich Park', drawn and engraved by Sutton Nicholls in 1723; all are open trestle post mills. A view of Greenwich by C. du Box in 1731 with the Isle of Dogs in the background shows 10, and there are several references in insurance policy registers and sale notices to the 'tenth mill'. The first map to show the mill is Jóel Gascoigne's of 1703, entitled 'An actual Survey of the parish of St Dunstan, Stepney'.

The history of the mills is confused. Their ownership, and thus the names by which they were known, would have changed over the years. Some were known simultaneously by more than one name. Gascoigne's map names them in the following order, going from north to south:

(1) Brown's Mill.
(2) Smith's Mill.
(3) Smith's Mill.
(4) Smith's Mill.
(5) Baker's Mill.
(6) Churn's Mill (also referred to as Chinn's, Chinner's or China Mill) called Tommy Tinker's.
(7) Ward's Mill called 'Theobalds'.

In a plan in a report of 1796 on improvements to the Port of London, eight mills are shown, and these are named as Nashe's, Lewis', Bingley's, Saunder's, Pye's, Kent's, Stiles' and 'Oil Houses' (?). The three shown on a map of 1835 are Theobalds, Tommy Tinker's and one called simply 'Flour Mill'.

The majority of the mills were post mills, but at least one, probably the tenth, was a smock mill; this features in a pencil sketch of 1827 in which its site appears to be a timber yard, judging from the lengths of sawn wood which surround it. The mill has common sails and a hand-winded cap similar in shape to some on German and Danish windmills. A circular structure surrounds the base.

A smock mill is also shown on a 19th century print of the mills, now in the Museum of London. This also has common sails but is winded by tailpole and has a stage around the lower part on vertical supports.

A fine watercolour painted by Miss H. J. J. West in 1830 depicts one of the post mills in greater detail than other pictures. It is a smallish affair with a single-storey roundhouse and common sails on which the canvas is unfurled, indicating the mill is at work. There is a decorative finial on the rear roof gable. A rope or chain braces the tailpole to the body of the mill, and a sack slide runs down one of the step strings. The lever on the tailpole is called a talthur, and is a feature of most post mills without automatic winding; it would have been hinged, enabling the miller to work it up and down and by means of a rod, connecting it to the steps, raise the latter off the ground, thus enabling the mill body to be turned more easily into the wind.

(1) first mill, Brown's

The 'two first windmills on Limehouse Wall' (this one and that sometime known as Smith's) were offered for sale in 1762. In April 1781 Captain George Russell of Rotherhithe and Richard Unwins of Cripplegate Buildings, apothecary, insured their house at Marsh Wall in the tenure of Mr Brown, miller, for £175. They also insured a timber windmill nearby for £25.

(2) second mill 1772 (presumably Smith's)

In May 1772 Ann Walford, who lived in Narrow Street, Limehouse, and is described as a corn chandler and slopseller, insured the windmill called the second mill for £240. Sarah Stile was tenant miller at the time.

(3) 'Middle' windmill

In October 1765 John Birch insured the 'Hazardous Building being the Middle windmill with a Ware House under it on the wall on the east side of Limehouse Reach . . . in the possession of John Pyer' for the sum of £200.

(4) '5th' windmill (Baker's in 1703)

In October 1804 Reverend William Tooke took out an insurance policy on his 'windmill called the 5th on Poplar Marsh Wall £300 on the standing and going gears, millstones, wire machines and dressing mills therein £150. On his corn windmill only, called the 6th on PMW £200, on the standing and going gears, millstones, wire machines and dressing machines £100 and granary near £100. All brick and timber built except where otherwise mentioned. Warranted no steam engine in, adjoining or communicating with the said mills'.

(5) Sixth windmill

As well as the above, John Birch insured the house and warehouse belonging to the 'sixth' windmill. In 1805 the mill, if the same one, was owned or tenanted, along with the '5th' mill, by Rev William Tooke. The document reads 'On his corn windmill only, called the 6th on PMW £200, on the standing and going gears millstones, wire machines and dressing machines £100 and granary near £100'.

(6) Chinn's (seventh mill?)

In 1739 Ann Price and Elizabeth Coates of Wapping leased to John Mitchell, lighterman, of Stepney, for the sum of £12 per year 'the windmill called the seventh mill or Chinner's on Poplar marshland or wall facing the Thames', the lease quaintly describing something of the fittings and machinery: 'The windmill . . . with the culling runner 12″ thick 4′ 3 high, the culling bed stone 7″ thick 4′3″ high, the peek-runner ½″thick and 5′2″ over the bedstone very much wore but 1 inch thick, brike on the neck for the shaffle, 2 top braces, a gudgeon brake, and six small box braces'.

On Gascoigne, Chinn's is the sixth mill; either someone is mistaken as to its location or another mill had been built and was now the northerly of the mills (that a mill is given a number, *ie* the 'sixth' or the 'eighth', is not necessarily a clue to its identity, since the order may have been changed from time to time by the disappearance of some mills and the appearance of new ones. It is assumed that they were counted from north to south).

(7) Ward's Mill

All we know about this mill apart from its name is that at some point the lower part was converted into a beerhouse, becoming known as 'The Windmill'.

(8) 'Tenth' mill

In August 1779 John Beck insured a smock windmill in an uncompleted state and separate from the other mills in his ownership for £400. Perhaps it was the same as that advertised for sale in the *Ipswich Journal* on 28 August 1784:

'To be sold by auction by Mr Chapman (on the premises by order of the proprietor) on Tuesday 31st inst. at 12 o'clock in 3 lots: A valuable leasehold estate, known by the Tenth Mill, on the river, situate on Poplar Walls 290 feet near the River Thames, consisting of a capital smock windmill (the compleatest in the kingdom) for grinding corn, which has 2 pairs French stones, a dressing mill, works by a horse[?], with sack tackle for unloading barges; also large and convenient granaries, which will hold 1,000 quarters; a convenient dwelling house, with some cottages, . . . at the same time will be sold, a large mill-boat, with sails, a skiff etc . . .'. It is probable that this is the mill shown in the illustration of 1827.

The following mills cannot be identified. John Beck insured a brick and timber windmill, along with adjoining granaries, for £1,300 in 1778. Greenwood's map of 1824-6 has two mills near Millwall Pier: one 40 yards to the north of it, the other 150 yards to the north and near the Catholic Church. One chronicler notes that a mill stood on the river bank opposite Strafford Street.

Only three mills stood by 1835. Ten years later a map of the Isle of Dogs published in the *Illustrated London News* showed only two, and one of 1859 has none at all. In 1853 B. H. Cowper, writing in his *History of Millwall*, stated that only one mill was still standing, on the premises of a Mr Weston. He described its remains as an octagonal brick building of three floors, which indicates either a tower mill or the base of a smock mill. It had recently ceased grinding by steam, and still retained some of its machinery. The foundations of two or three other mills could still be traced.

One of the mills appears to have found its way into the writings of Charles Dickens. On page 63 of *Our Mutual Friend* is described the home of Gaffer Hexam, who makes a living fishing dead bodies from the Thames:

'The low building had the look of having once been a mill. There was a rotten wart of wood upon its forehead that seemed to indicate where the sails had been . . . The boy lifted the latch of the door, and they passed at once into a low circular room . . . there was a wooden bunk or berth in a corner, and in another corner a wooden stair leading above — so clumsy and steep that it was little better than a ladder . . . The roof of the room was not plastered, but was formed on the flooring of the room above. This, being very old, knotted, seamed, and beamed, gave a lowering aspect to the chamber; and roof, and walls, and floor, alike abounding in old smears of flour, red-lead (or some such stain which it had probably acquired in warehousing), and damp, alike had a look of decomposition'. This is certainly an evocative, and not inaccurate, description of the interior of a windmill. If it was a real mill it is most likely to have stood at Millwall.

Elsewhere in Poplar, a mill is shown on the 1844 Ordnance Survey map in the vicinity of Carmen Street. In the same year what must be the same mill is mentioned in the *London Gazette*; William Hazel, miller, resided at a mill in Kerby Street, off East India Road, on the west side of the railway line.

A windmill also stood at Blackwall, and is shown in a pencil sketch of the riverside there, dated 1840. It is either a smock or tower mill with common sails and a cap similar to that on the Millwall smock mill. A bulge at the rear of the latter may represent housing for equipment for winding it by hand.

RUISLIP

Early mill. 099876

A windmill and a watermill, valued together at 40s, existed at Ruislip in 1294 according to a survey of the parish. The former stood at Wyndmyllhyll to the south of the village. There are no further references, but the site continues to be known as Windmill Hill. On Doherty's map of 1750 the open common land which covered the hill is divided into Great and Little Windmill Fields. The map shows a small patch of pasture, thought to have been the site of the mill, to which a narrow strip of green, perhaps once the approach track, leads from an eastward extension of Brickwall Lane. It is said that the building's foundations were uncovered when the first houses were built at the top of the hill.

An old map is said to show a windmill on the common at Northwood.

ST CLEMENT DANES

17th century mill. 311812

In 1615 a tailor named Robert Baker paid rent for grazing ground 'near the windmill in the Strand'. This mill stood near St Clement Dane's Church, at the top of Milford Lane. It is shown in an engraving by Visscher as a typical open-trestle post mill of the period. R. Thurston Hopkins believed it to have disappeared during the reign of Elizabeth I.

ST GEORGE HANOVER SQUARE

(1) 17th century mill 287813 (original site)
(2) St George's Hospital. Standing 18th century. 284798

Ogilby's *Britannia* (1675) depicts a windmill at the top of Bond Street; later moved to Mill Hill Place, off Oxford Steet.

In the mid-18th century a millwright named Thomas Yeoman built a small windmill on the roof of St George's Hospital which, like the similar mill at Newgate Prison, operated ventilators. A diagram in Abraham Rees' *Cyclopedia* shows it took the form of a small open-framed, hollow-post mill mounted on a trestle structure and winded by a weathervane. The ventilation system was devised by Stephen Hales, as for that at Newgate.

ST JAMES

C17 mill. 296807

The famous Windmill Theatre derives its name, indirectly, from a windmill. Robert Baker built the mill in the early 17th century on the northern portion of a piece of land he had acquired from John Golightly. The first map to show it was surveyed by Richard Newcourt between 1643 and 1647 and engraved by Faithorne in 1658. The mill then stood on the western extremity of London, with open country to its north and west, to the south of the junction between Great Windmill Street, as it came to be called, and Brewer Street, close to the spot where the modern Haymarket now joins Coventry Street. Great Windmill Street was originally just a track leading from Tottenham Court Road to the windpump in Rathbone Place, Soho.

According to C. L. Kingsford, in his *Early History of Picadilly, Leicester Square and Soho*, 'the windmill itself, which stood to the west of the north end of the street, was fitted with staves and strongly built of brick covered with tile, with large lofts and corn. The residential part of the structure was however old and decayed, consisting of a cellar with three rooms over it, one above the other'. This suggests the mill was a tower mill, or at any rate a smock with a substantial brick base. Although Newcourt's map shows a post mill, it is at least possible that the mill was rebuilt as a smock or tower at some time after the mid-17th century.

The area was developed comparatively early, and there is no evidence that the windmill existed beyond the middle of the 17th century. A few of the houses nearby, shown on Newcourt's map, survived into the 1880s.

ST MARY LE STRAND

Possible mill near Drury Lane Theatre.

The name 'Windmill Tavern', by which a public house on the site of the present Drury Lane Theatre was known, may indicate the former existence of a windmill.

ST SEPULCHRE

Newgate Prison, standing 18th century. 318185

At Newgate Prison there was for a time a good example of new uses for windmills in urban areas.

In October 1750 the Corporation of London, alarmed by the spread of disease in the prison, appointed a committee. It was decided that the trouble lay with the ventilation system and, at the recommendation of a Norfolk parson, Stephen Hales, a new system was installed, designed by Hales himself and operated by a windmill erected on the roof of the prison, where it would have less difficulty in catching the wind. The millwright was a man named Cowper, from Penny Fields in Poplar. The project was not completed until 17 April 1752, partly because many of those employed in fitting the new system became ill from the foul air it was designed to dispel.

The structure stood about 15 ft high and resembled a hollow-post mill. The sails were in the form of an eight-bladed fantail. A crank on the windshaft was connected by a shaft passing down through the hollow centre of the post, to a lever 14 ft long, which was pivoted to raise one set of ventilators while depressing a second. The mill obviously had to respond to a light wind; otherwise ventilators were operated by hand. A detailed description appeared in the *Gentlemen's Magazine* in 1753. An extract includes this observation:

'As an angle of 55 degrees are found, both by mathematical calculations and by experience, with a small windmill placed before the nose of my ventilators, to be the degree of weathering, as millwrights called it, or the angle or oblique position of the sail to the wind, for obtaining the greater force, so an angle near that, viz. of 60 degrees, has been found by experience of the Rt. Hon. the Earl of Northumberland, as he informed me, to do very well in such small mills. But in common mills, with very long arms, the obliquity of the sails ought to be less; else their great velocity will, in strong winds, cause a counteracting force at the back of the sails. In order to avoid this inconvenience, Mr Cowper very rightly proposed, not to fill up the whole of the space with the sails, but to leave a void space of about 6 inches breadth between the sails, as is represented by the sails mm, that the direct current of the wind, as it passed through there, might carry along with it, and give a turn to the course of the wind, which else, being drove obliquely from the face of the preceding sail, would thereby be drove to act on the base of the following sail, and thereby aboute the force and retard the motion of the mill'.

Just how successful the new system was in practice we do not know, but it was evidently an improvement on the previous one. Dr Hales reported in 1753 that the number of deaths had fallen off in the four months during which it had been in operation. It was employed up to the time of the old prison's demoliton in 1778, for the mill is shown on a plan of 1777 and is also depicted in Harrison's *History f London* in that year.

SHADWELL

Standing 1824-26. 354807

Greenwood's map of London, surveyed between 1824 and 1826, shows a windmill by the east side of Shadwell basin.

SHEPPERTON

(1) Early mill.
(2) C16 mill near Walton Bridge. 090658 (approx.)

A windmill was built in the manor of Halliford in either 1381 or 1382; it was probably intended to replace one which earlier stood in Upper Halliford. The later mill survived for about twenty years. Research carried out by a student of Dr John Langdon of the University of Alberta suggests that the mill's demise was due to the Black Death.

Its construction in 1381-2 cost £39 5s 9d. The timber for the main structure came from London and Greenford; other components, including two brasses weighing 72 pounds each and costing 15s, were brought from Wandsworth. As well as the estrich boards for cladding the mill, 20 made from elm were also required. A cable used to lift components cost 3s 6d. A man was employed to build hedges and a ditch around the mill with two gates to keep the latter clear. The mill was valued for the year following its construction at 40s 4d and was leased to William Mullen-Ward. The records also state that the post had been weakened by a flood. In 1383-4 the mill was of no value because it had no tenant, he having died from the plague or fled the area. However, shortly afterwards John Hessendene took out a seven-year lease. The value of the mill had now decreased to 26s 8d. The cost of fitting a new 'batell' was 22s 1d. Hessendene was still there in 1388-9 but, after the expiry of the lease, no-one else took it on and in 1398-9 it was reported derelict and unoccupied.

A windmill was standing in the parish in 1597 when it was left by will with the proviso that it should not be moved. Walton Bridge Green and Walton Lane (formerly known by the names of Windmill Common and Windmill Lane respectively) mark its approximate site.

SHOREDITCH

Orsman Road, Hoxton. 335836

Greenwood's map of 1824-6 depicts a mill 200 yards along Canal Road, now Orsman Road, which runs west from a point near the library in Kingsland Road. No other map marks it, suggesting its existence was brief, but it stood long enough to give its name to nearby Mill Row. The symbol is of a post mill, although the mill was not in fact of this type, as a sale notice in the *London Gazette* of 13 October 1829 tells us:

'[It was] A powerful smock windmill, erected within these few years in the best manner, being brick-built . . . [it] contains six floors, fitted up with two pairs of French Burr stones (but is capable of driving six pairs), sack tackle, and all necessary gears . . . and an iron foundry'.

The mill was then held under a lease granted to the late Charles Harland, of which 47 years had expired. The description of brick-built tower mills as smock mills is common in sale notices of the period. The mill had been sold earlier, but was now resold in accordance with an order made in a High Court case between Harland's heir and one Bilton.

An illustration of Baumes, a large house of 1650 on the west side of Kingsland Road, shows a tall tower mill with a stage about half way up, and a domed cap with a lofty fantail, standing in the background. It is likely that this is the Canal Road mill; the sale notices suggest a large and powerful affair. The resemblance to the White Lead Mills at nearby Islington, down to the bowsprit in front of the sails, is striking. Walker's mills had clearly captured the public imagination, instilling in some entrepreneurs a desire to build a mill as large and as technologically advanced.

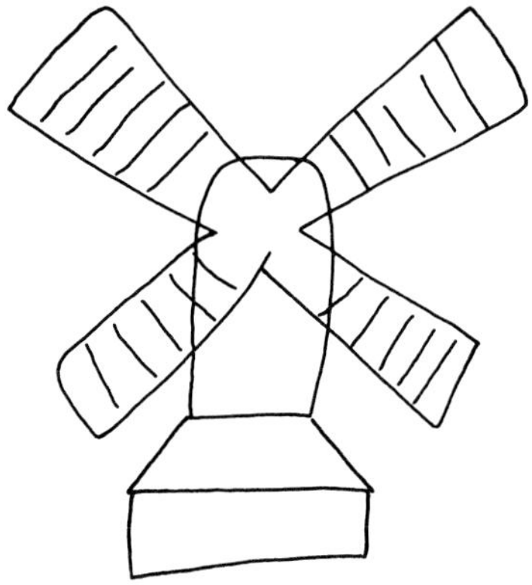

Post mill at Norwood, from Moses Glover's map of the Isleworth Hundred, 1635. (PB)

SOHO, RATHBONE PLACE

18th century mill. 294819

During the 18th century there stood in the centre of Soho Square an elaborate ornamental fountain worked by a windmill mounted on a tower at the north end of Charlotte Street. The water was pumped from a pond in the street, which was fed by a spring originally in open ground but incorporated within the cellar of a house, when the area was developed. The 'mill' could only operate when the pond was full, so as to save the works from unnecessary wear. According to one author it also ground corn.

In 1832 the mill was remembered by the sculptor Joseph Nollekens, which suggests it was standing in 1787. Nollekens recalled that the owner of the land on which it stood, perhaps also its operator, charged the public a halfpenny each for the privilege of walking in his grounds.

The mill is illustrated in a print in Strype's 1721 edition of John Stowe's *Survey Of London*. It was a small tower mill with ogee cap and sails of indeterminate type, reminiscent of the 1712 illustration of Clerkenwell windpump. The rolling fields in the background show that it then stood on the fringes of London.

SOUTHALL/NORWOOD

(1) Norwood. Standing 16th century and later.
(2) Post mill near Grand Junction Canal. 144797
(3) Smock mill, near Uxbridge Road. 143809

In 1596 the executors of Anne, Lady Dacre, reserved for themselves a watermill and gardens in Norwood and a windmill and one acre of land, both copyhold of Norwood Manor. The mill may be the same as that standing in the manor of Southall in 1433 and 1496, since in a conveyance of the manor in 1597 none is mentioned.

After that date there is no firm evidence of a windmill in Norwood until the 18th century, although a mill of unknown type was standing in 1611. This, along with the overshot watermill at Northcott, was leased in 1676 to a Mr Hamton. Moses Glover's map of the Isleworth Hundred, published in 1635, shows a post mill beside the river at Norwood Green; this appears to have a roundhouse, which would have been the earliest for which there is evidence in Britain, although one of the other post mills on the map also possesses something like a roundhouse, and has been so described, when it is just as likely to be a representation of the piers supporting the trestle.

The mill which stood near the boundary between Southall and Hanwell is first recorded in 1721 as the scene of a robbery, reported in the *London Gazette* of 19-22

August. It was probably the same mill which was among the properties owned by the Cooke family, lords of the Manor of Hayes, and sold by them in 1770. And the following intriguing reference is to be found in the Hanwell Vestry Minutes of 20 September 1800:
'The Rector reported that the nuisance from the mill having continued, and indeed being aggravated, he had preferred a bill of Indictment against Messrs Wallace and Hawes'.

In 1806 the windmill, along with Windmill Farm, formed part of the Osterley estate. It stood by the Grand Junction Canal and was presumably the one painted in the same year, in somewhat idealised fashion, by William Turner, who depicts it as a post mill with a roundhouse. It was still standing in 1821 when it is shown in a plan, drawn up by one Lediard, which accompanies the Valuation List for the Norwood Precinct. It is noted as near the Three Bridges, where the canal is carried over the railway and the road over the canal. The mill was owned by the Earl of Jersey and occupied by Jeremiah Sibley, and was worth £15. It is possible that the structure or its remains were demolished in 1855 when the railway was built from Southall to Brentford.

By the 1820s, when the post mill is last heard of, another windmill had made its appearance. A smock mill, it stood close to the overshot watermill with which it has frequently been confused by local historians. The site was about half a mile from the main Uxbridge road and about 50 yards to the west of the road near a prefabricated housing estate.

Writing in the 1960s, D. F. Kiddle stated that a number of illustrations of the mill were in existence, but research has failed to find them. R. Symonds was the miller in 1855 and 1862 and the mill is last known to have worked in 1886 when an R. P. Symond is listed in directories.

The last map to show the mill is the Ordnance Survey of 1877. In 1869 an elderly man wrote to a local newspaper to say that as a boy he had helped to demolish it towards the end of the previous century. Today the approximate site is marked, most conveniently, by an Ordnance Survey triangulation station. A few bricks from the base of the mill remained into the present century, and a piece of wood said to have come from the upright shaft is preserved at Southall public library.

STANMORE

(1) Early mill.
(2) Early mill.
(3) C16 mill.
(4) Early mill. 168940

A windmill existed at Stanmore in 1352. The mediaeval chronicler Walsingham records that John de Maryns, Abbot of St Albans, whose abbacy was from 1235 to 1260, built a windmill and a manor house in what later became Great Stanmore.

In 1547, following the dissolution of the monasteries, a windmill which had previously belong to St Bartholomew's Priory was granted to Pedro de Gamboa along with two horse mills. 'A little house called a mill house' was claimed by a customary tenant in 1665; by then the mill itself had disappeared.

The cellarer of St Bartholomew's received 46s a year from a windmill at Grimsditch, in what later became Little Stanmore, in 1306. A mill house was among the appurtenances of Canons leased to Hugh Losse in 1543 and a mill among those surrendered to Sir Thomas Lake in 1604. This mill was said in 1680 to have stood near the boundary with Great Stanmore, on the crest of Brockley Hill.

STEPNEY

(1) 14th century mill.
(2) 18th century mill, Ratcliffe.

In 1324-5 there was a windmill in Pomfret Manor on Stepney Marsh, and one of unidentified type, which may have been the same, stood there in 1342. It may have fallen into disrepair because of changes in land-usage and the flooding of the marsh.

In 1712-13 Roger Goodcheape insured his house 'near the windmill in Brook Street in Ratcliff in the parish of Stepney'.

SUNBURY

(1) Early mill.
(2) Standing 1591 and 1603, site unknown.
(3) C17 and 18th century mill. 099699
(4) Windmill Road. Survived into early C20. 091701

A valuation carried out in 1311 of the rectory and vicarage of Sunbury states the tithes from a windmill to be worth 3s 4d. A windmill is again recorded in the area in 1591 and 1603.

Some time between 1676 and 1683 Henry Biddle, a millwright from Easthampstead in Berkshire, took over a copyhold windmill on Sunbury Common from Katharine Cotton, a widow, surrendering it in 1683 to Henry Thorniworke. On 1 October 1689 Thorniworke transferred the mill to his son, also Henry, from whom it passed in 1696 to George Cromwell. The Thorniworkes probably continued as tenant millers since Cromwell was not a miller himself and lived in Hampton. In 1704 Henry Thorniworke, miller, is shown as a copyhold tenant at a rent of £1 13s 4d. In or shortly before 1705 the mill was destroyed by fire.

As local historian Mr Kenneth Heselton has pointed out, the mill appears to have been the most expensive property in the Manor. The rent was almost as much as that for genteel residences like Hawke House, the home of Admiral Hawke, which still stands. Rent continued to be paid for the mill after it had been burnt down, as well as for a ruinous and uninhabited cottage nearby. The mill must have been exceptionally profitable as it was soon rebuillt. In 1712 it was let to William Cooper, miller, for £12 per year. It is shown on a 1722 map of Sunbury Manor, standing to the south of the Staines Road and east of the western boundary of the Manor, which ran due north from Laytons Lane to the main road. The site was probably on the line now followed by the motorway.

By 1819 another windmill had been erected on the west side of what is now Windmill Road. This mill was a large smock, with room for three pairs of stones, a flour machine, a smutter and a bolter. It stood on a shallow brick base six or seven feet high. There was a fantail and probably patent or spring sails. Also on the property were a large storehouse capable of holding 100 sacks of flour, a stable, two cart sheds, a barn, a house for the miller with four bedrooms and two bedsitting rooms, a washhouse, bakehouse and garden.

In July 1847 it was auctioned at the Flower Pot Inn, and it is from the sale notice in the *Sussex Advertiser* that these details are gleaned. The miller in 1852 was James Fry, a local man who in the same year was listed as an insolvent debtor (*London Gazette*, 13 April). Nonetheless, the mill went on working until at least the late 1880s and may not have finally stopped until the 1900s. Millers after Fry were W. Wells (1867), Thomas Winter (1870), Stephen Laslett (1874, 1878), and a Mr Weekes or Wells in 1886.

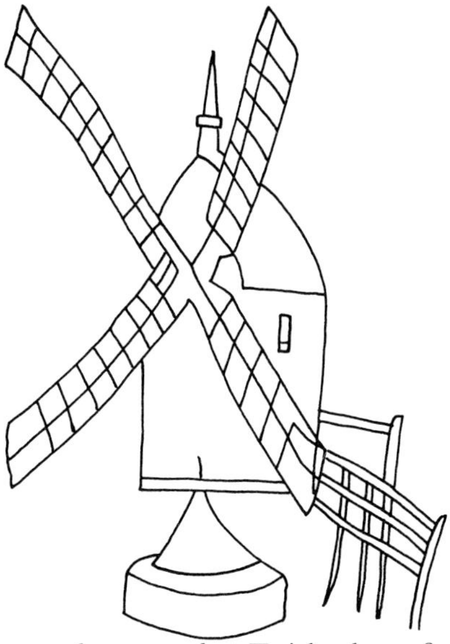

Post mill which may have stood at Twickenham, from Moses Glover's map of the Isleworth Hundred, 1635. (PB)

At some time the wooden smock tower was pulled down and much of the timber from it incorporated in a structure which was erected on the base to form a barn. This survived until c1950. Today the grassed-over site at the end of Lincoln Way is owned by Thames Water.

The mill is still remembered in the locality, principally as a stump, although in 1991 one elderly resident could still recall playing in it as a child in the early 1900s, before the main part was demolished. Today the Windmill Road area is a hive of local industry, in which the name of at least one firm, Windmill Plastics, commemmorates the mill's existence.

TWICKENHAM

(1) Early mill.
(2) C17 mill (possibly in Surrey).
(3) Near Hospital Bridge Road. 138729 (approx.)
(4) Possible windmill in Crane Park, the tower of which stands today. 128728
(5) Pumping mill, standing in early C18.
(6) At Fulwell watermill. 144728

In the 14th century a windmill existed in Isleworth manor, which included parts of Twickenham. Moses Glover's map of the Isleworth Hundred, drawn up in 1635, depicts three post mills, one of which is on the Surrey side of the river at Richmond. In the lower right hand corner of the map is a rather delightful illustration of another, whose location is not clear. Beside it a note reads 'This mill s(t)andeth hereto without ye parke'. There is a curious spire-like structure on the roof. The third mill was near 'Whittne Bridge', close to the junction between the present Percy Road and Hospital Bridge Road, about fifty yards east of the roundabout. It disappeared some time between 1675 and 1743.

There was a windmill at Whitton in 1784, for a land purchase agreement of January that year mentions 'land and that said parcel of waste or ground upon Hounslow Heath lying and being near the windmill at Whitton'. It may have been that erected at Fulwell watermill; no more is heard of it after 1799.

In 1722 the eccentric Lord Archibald Campbell, Earl and Viscount of Islay, acquired some 400 acres of land on Hounslow Heath, adjoining the hamlet of Whitton, from the Crown. He set about turning this into an estate, divided into two sections, separated by the road leading north to Hounslow. The western section was surrounded by a drainage and irrigation ditch feeding a number of streams, the flow of which was controlled artificially by a 'wind machine'.

There the subject of Twickenham windmills might be closed, but for an impressive brick structure in Crane Park, Whitton, whose origin has never been satisfactorily established.

The building certainly has the appearance of a windmill tower. A note by M. T. Mason in the Simmons Collection states 'There was a tower mill at Twickenham, tall and black with 4 sails'. 'Tall and black' is certainly an accurate description of the structure in Crane Park. It could have been the windmill standing at Whitton in 1784, but this conflicts with the evidence of the 1828 datestone, now no longer visible. According to G. J. Aungier in his *History and Antiquities of Syon Monastery, Isleworth and Hounslow* (1840), the water was pumped to the wheels by 'a curious pump, worked by wind sails', which was capable of raising from 30 to 50 tons of water a minute. It is possible that Aungier is referring to the Crane Park tower; if so, he has got the location of the mill wrong, for he states it to be 'about two miles to the south-west of Hounslow', which would place it at the western edge of the complex, near Baber Bridge which is in Bedfont. Perhaps the strongest evidence for the 'windmill theory' is the presence of two rows of square openings in the brickwork, now mostly filled in. Judging from their position and distance apart they could have held the horizontal members, and footings for the diagonal supports, of a stage of the kind found on some tall tower mills. Windmills may not have been the only buildings to have such structures but the existence of one here does tend to make the windmill theory more plausible.

In the 19th century the structure was known as the 'Mill Head Tower', suggesting its purpose was to store water to provide a constant head for the waterwheels which drove the machinery at the gunpowder works. However, according to a former employee at the works, the level of water was kept up by means of a hydraulic pump.

The building has had a number of different uses: it has been a watch tower for the gunpowder mills and a workshop for making and repairing sieves, and at one time also housed machinery connected with the process of pelleting (compressing gunpowder into cartridges). At present it is being converted into a study centre for the surrounding nature reserve. For some time it was thought to be a shot tower, and sports a plaque describing it as such. It is unlikely to have been anything of the sort. It is not the right height — most shot towers were 150 ft high or more, whereas the Crane Park tower is only 80 ft — besides which the furnace which would have been required for melting the lead would have been a safety hazard in a gunpowder works. If the structure was ever a windmill, it cannot have been one for long as otherwise there would be more evidence.

WESTMINSTER

(1) Early mill.
(2) Millbank.
(3) Near Charing Cross, standing 1696.
(4) Proposed pumping mill, Tyburn.

A windmill at Eye, the demesne farm for Westminster Abbey in mediaeval times, is mentioned in the Abbey records from the 1270s to 1305-6.

LEFT: *The White Lead Mill, Millbank*, from a pencil sketch by Hendrick de Court, c1800. (PB) RIGHT: Windmill near Charing Cross, 1696. (PB)

An unfinished pencil sketch c1800 by a Dutch artist depicts a smock mill, called 'The White Lead Mill, Millbank', close to the water's edge, with common sails and a domed cap winded by hand. The weatherbeam and the front ends of the sheers protrude beyond the circle of the cap roof – a feature usually associated with mills in Lincolnshire and the North-West – as do the horizontal members of the cap frame.

A print of 1696 by Robert Morden and Philip Lea, entitled 'A Long View of London and Westminster taken from Southwark', shows what appears to be a post mill standing on another building near to Charing Cross and Northumberland House.

In 1592 there was a proposal for an engineer named Frederick Genebelli to erect a pumping windmill at Tyburn, but nothing seems to have come of it. Windpumps seem to have been a speciality of Genebelli's; he built one on marshland near Rye in Sussex.

WHITECHAPEL

(1) C17 mill, Fieldgate Street. 343816
(2) Fieldgate Street. 342814
(3) Standing 1806 near George Inn in Commercial Road.

A mill stood at Whitechapel in 1642-3, during the Civil War, when the Parliamentarians who controlled London built extensive fortifications nearby. It survived the war and is shown on a map dated 1658 as an open-trestle post mill, standing 200 yards east of St Mary's Church at the west end of the present Fieldgate Street, which at that time marked the end of the village of Whitechapel, the mill standing at the edge of open fields – to which a gate gave access.

A mill is shown at more or less the same position on Sutton Nicholl's view of London and Southwark in 1724, but this is depicted as a tower mill. It stands to the north of the Tower near St Mary Whitechapel. A tower mill is also shown on a print regarded by Farries and Mason as of doubtful authenticity. This being before the advent of the fantail, its cap is winded by a tailpole.

In 1752 John Cardell of Mile End Old Town, clothworker, insured a 'timber building being a windmill in ye Tenter Ground in ye Parish of St Mary Whitechapel being a little west from ye church and ye only windmill thereabouts in his own possession'. If Cardell's mill was built of wood it cannot have been a tower mill. However, it is just possible that it was a smock mill, From its position it could not have been the post mill standing in the 18th century, unless it had been moved.

The Fieldgate Street mill is last heard of in 1807, when it was offered for sale along with the nearby Black Horse public house. In the previous year another windmill stood near the George Inn on the north side of Commercial Road, near the end of the former Daran's Row.

WILLESDEN

(1) Early mill.
(2) Dudden Hill. 218854 (approx.)
(3) Shoot-Up Hill. 245848

A windmill had been erected here by 1295, to the north of the Sherrick Brook on an estate conveyed by William of Breadstreet to John of Middleton. By 1365 it was in such a ruinous condition that repair was thought uneconomical and the mortgagee, Thomas Frowyk, was given permission to take the timber from it for his own use. The mill was not replaced, and Willesden was without a windmill until 1616 when William Grey, a lessee of Francis Roberts, built one in Dudden Hill Field, at the point where Dudden Hill Lane entered it. This mill was held by Robert Paltock in 1698; he was later succeeded by John Nelson who built a house and two cottages there. Nelson sold it to a Mr Hardcastle of Holborn, from whom it passed to Edmund Franklin in 1727. The last recorded tenant is William Kilby in 1765. The mill had gone, or was derelict by 1817, when a local rate directory refers to land among the property of James Hall, known as Mill Piece, 'Where the mill and millhouse and two tenements formerly stood, now fallen to decay'. Local historian B. C. Dexter, writing in a local paper during the 1890s, stated that a few scant traces survived until obliterated when the track of the Midland and South Western Junction Railway was laid in the 1840s. The mill is shown as a post mill on a map of road repairs in 1749, with the sails projected further forward of the body than would have been the case.

Some time between 1784 and 1803 Isaac Ennos, a tenant of Mapesbury Hall, built a smock mill on Shoot-Up Hill. It became known as Kilburn Mill, despite being some distance away. In 1798 Ennos leased it to William Bonnington. Bonnington had the mill insured against fire for £1,000 — £600 for the mill structure and £400 for the machinery within — in 1806. In common with many others the agreement included the clause that no steam engine should be installed, sparks from the boiler being a fire hazard.

Ennos was succeeded around 1832 by William Hale, who in 1851 employed five men at the mill, an indication that the businews was prosperous. In 1863 the mill was largely destroyed by fire, which led to the establishment of the Kilburn and District Volunteer

Fire Brigade. The need for a locally-based fire fighting service was demonstrated when the nearest engine, that stationed at Paddington, arrived too late. Among the crowd which watched the mill burn was the then Prince of Wales, later Edward VII; he took an interest in fire-fighting and owned a fireman's uniform.

The conflagration was popularly believed to have been due to the mill's sails turning too fast in a strong wind, the resulting friction generating enough heat to set the machinery ablaze. This kind of thing did happen to windmills occasionally.

The usefulness of the new organisation was demonstrated in 1872, when they put out a fire at the steam mill which succeeded the windmill.

Post Mill at Dudden Hill, Willesden, from an 18th century plan of road repairs. (PB)

Instead of rebuilding the windmill the Hales had decided to invest their capital in a more modern power source. The steam mill was built in 1867, four years after the windmill was burnt; that the project took this long is indicative of the severe blow to the family's fortunes, for the fire also destroyed their house. Ironically, in view of its youth and modernity, the new building came to be known as the 'Old Mill', leading to some confusion between it and the surviving portion of the windmill, which stood close by and to which the name was also applied. Equally ironically, it was to have a short life, ceasing work in 1891 and being demolished in 1898. In fact the windmill, whose last remains were not cleared away untl the 1900s, outlasted it. The steam mill was a rather ugly building, wider at one end than at the other, which made the roof an odd shape, and its disappearance was not lamented.

In a letter to the *Middlesex Quarterly* in 1954 an elderly lady remembered playing as a child in the charred remains of the windmill, which she described as consisting of 'one wing with about six slats on it'. She continued, 'My brothers and I used to swing on it, but being old and burnt it soon fell to pieces'. These remains collapsed in 1903 and any surviving traces were obliterated when a house was built between 1905 and 1910.

There exist several photographs, and two pencil sketches in the Grange Museum at Neasden depict the mill in its unmodified state, with common sails and manual winding gear.

The photographs show the mill from the front. The smock tower is broad and not dissimilar in proportions to that of Upminster mill in Essex. It is not clear whether there is any brickwork in the structure, as the lower part is hidden on the side facing the camera by a wooden building, set close against the mill, whose roof would have served as a stage (and therefore probably encircled the whole building). It may also have provided extra storage space and housed an office for the miller, as did similar structures found on smock and tower mills in Kent and Essex. There were four single-sided sails, one pair of commons and one pair of springs, and a fantail. Seen head-on the cap looks to be of the Norfolk type, shaped rather like an upside-down boat, though in the sketches of the 1820s, drawn from the side, it appears more like those of Lancashire windmills.

YEADING

Willow Tree Lane. Existence unconfirmed. 115821 (approx.)

In the 1950s M. T. Mason was told that a windmill once stood here, but despite making extensive enquiries he was unable to find any record.

Acton smock mill as painted by John Varley, 1835. (WG, UOM)

ABOVE: Acton smock mill by J. Gudet. (LBELS) BELOW: View of Chelsea Waterworks by J. Boydell, 1762, showing on the right what is probably the base of the windpump. (MOL)

ABOVE: The base of the New River Head windpump as it stands today. (GB) BELOW: The windpump at New River Head, Clerkenwell, from an 18th century painting. (MOL).

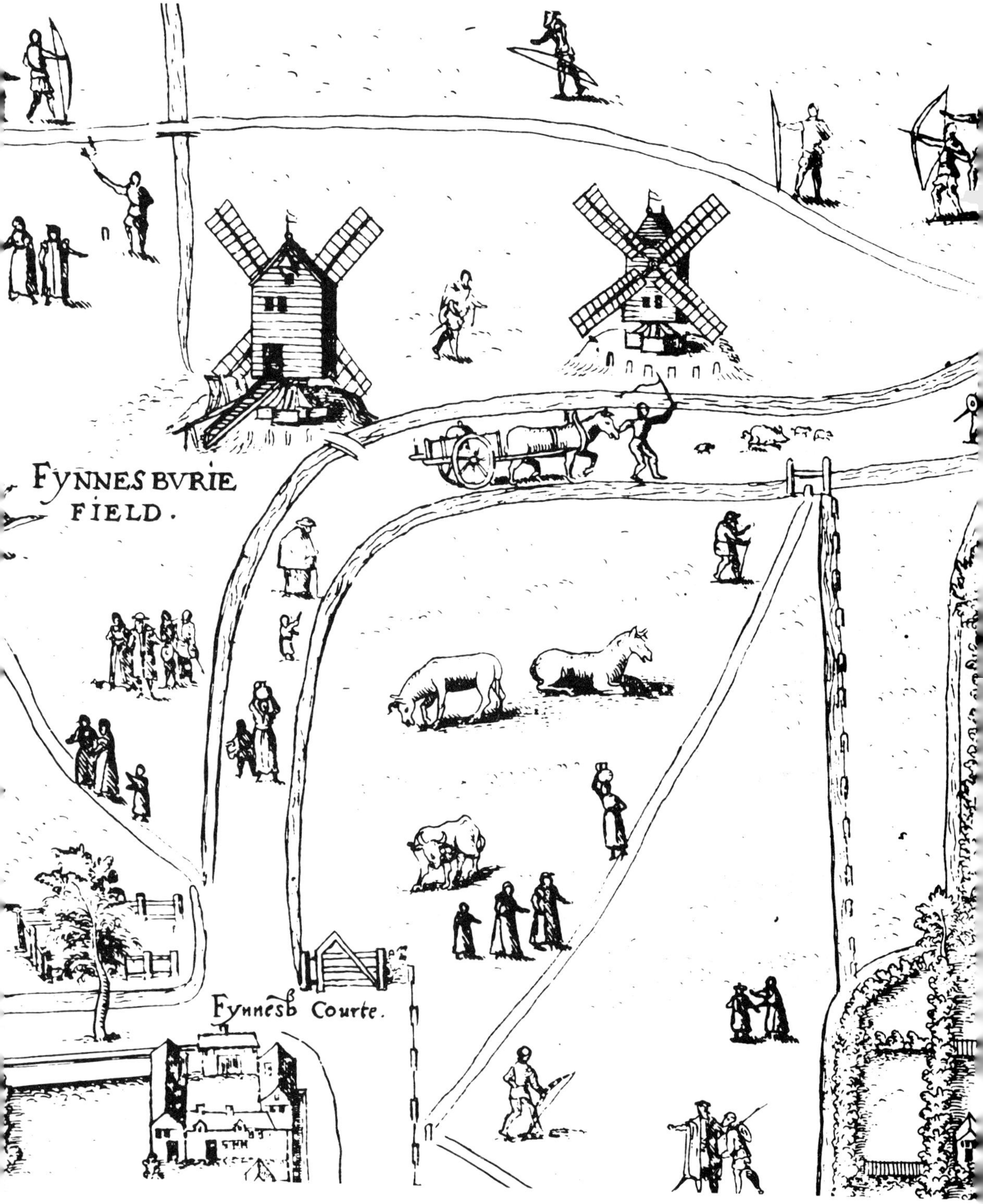

Two of the post mills at Finsbury: details from a map of 1560. (MOL)

ABOVE: Edmonton mill, from a 1905 postcard, (LBOEL) and BELOW: in extreme decay, c1920. (LBOEL)

LEFT: Enfield smock mill, late 19th century. This is the only photograph to show the mill with all four sails. Note the hideous Victorian house to its left. (LBOEL) RIGHT: Enfield, c1890. The ladder leaning against the cap suggests maintenance. (LBOEL) BELOW: Enfield, c1900. (LBOEL)

LEFT: Hampton smock mill: a retouched photograph of c1870. (MO) RIGHT: Hampton smock mill, from a late 19th century engraving. (LBRUT) BELOW: *High Ground in North-West London*: painting by Constable. (MOL)

ABOVE: The 'Brazil Mill', Hounslow, 1757. (OCR) BELOW: Heston Mill, Hounslow, 1884. (LBH)

ABOVE: The White Lead Mills, Islington: watercolour by Julius Caesar Ibbotson. The five-sailed mill is the one on the left. (MOL) BELOW: The same mills, by C. H. Matthews, 1835. (GLRO, M&P)

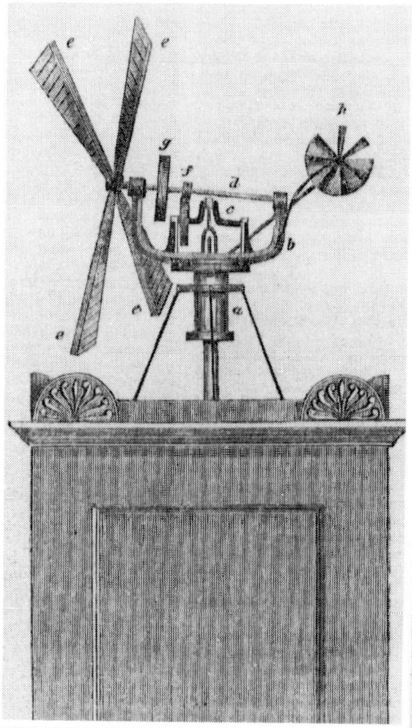

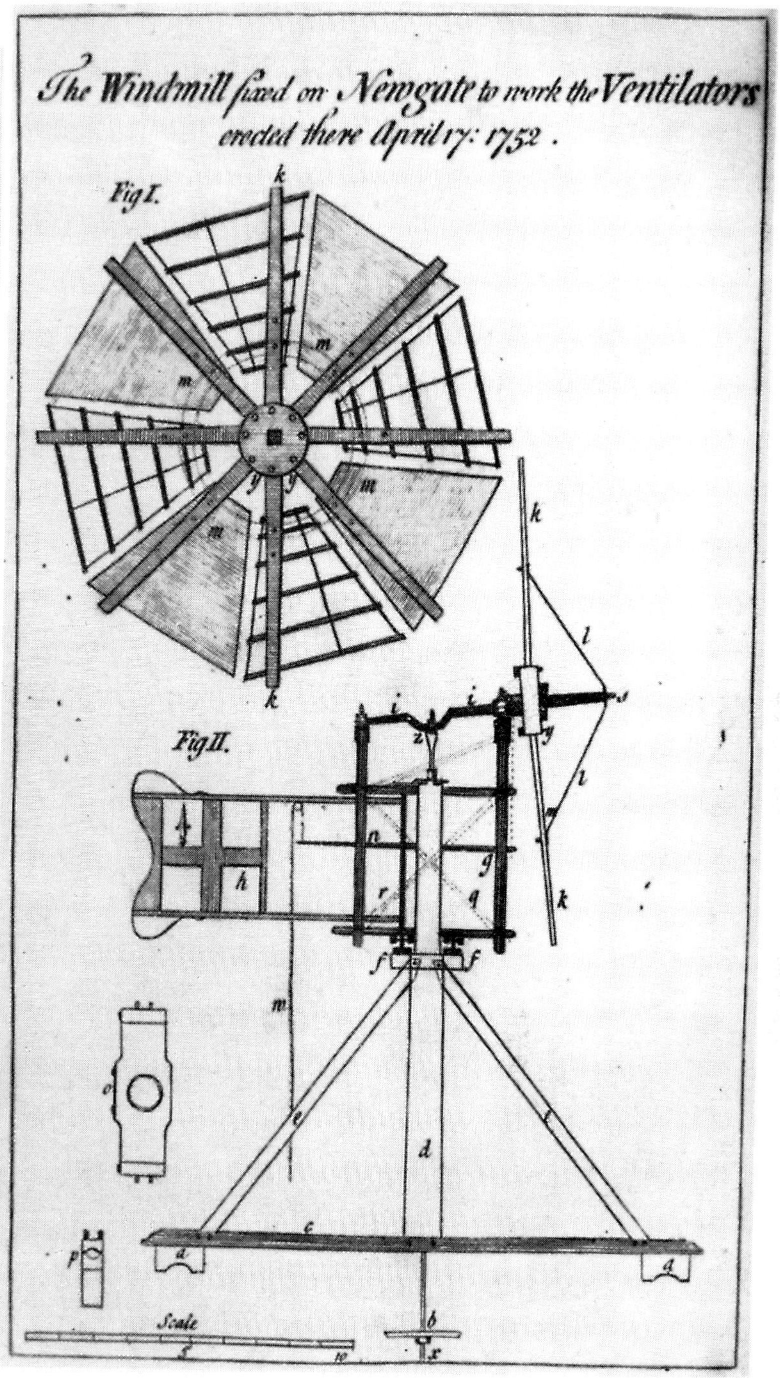

LEFT: One of the two windpumps at Islington Cattle Market, from the *Mechanics' Magazine* of 27 May 1837. (GL, COL) RIGHT: Diagram of the Newgate Prison windmill. (GL, COL) BELOW: The Newgate Prison ventilating windmill, 1777. (GL, COL)

ABOVE: Pinner smock mill c1870, shortly before its destruction by fire. (DRC) BELOW: Smock mill at Millwall, 1827. (LBTH, LHLA)

ABOVE: Hoxton tower mill, date unknown. (HAD) BELOW: The windmills at Millwall. (MOL)

LEFT: Shoot-Up Hill Mill, modernised apart from one pair of common sails. The photograph is undated but was probably taken not long before the mill's destruction by fire in the early 1860s. (CLA) BELOW: Shoot-Up Hill Mill, Kilburn, with common sails and manual winding gear, 1823. (CLA) RIGHT: The possible windmill tower in Crane Park, Twickenham. (LBRUT)

ABOVE: The pumping mill at Rathbone Place, Soho. (GL, COL)
BELOW: *Windmill and Lock* at the Grand Junction Canal, Southall, by William Turner, 1815. (BM)

FOR THOSE WHO MIND ABOUT MILLS

The windmills of Middlesex may have passed into history, while little of the watermills remains, but in other parts of the country many may still be found, in varying states of preservation. Those in a derelict condition are obviously at risk, while restored mills need constant care and attention if they are to survive. If this book has kindled in the reader an interest in the preservation of these beautiful structures, or in finding out more about them, they might like to join the Wind- and- Watermill Section of the Society for the Protection of Ancient Buildings, which is the national body concerned with the conservation and study of mills. It organises lectures, publishes a newsletter four times a year, and holds annual Windmill and Watermill Meetings in London. On the preservation side it campaigns in defence of mills threatened with demolition or unsympathetic conversion to other uses, taking up the matter with local authorities and giving evidence at public enquiries.

For further details contact the Section Administrator at SPAB, 37 Spital Square, London E1 6DY.

Glossary

Bridge Tree — Beam supporting the spindle on which a runner stone is mounted. It can be lowered or raised to increase or decrease the gap between the runner stone and the bedstone and so vary the texture of the meal.

Buck — Term of Suffolk origin sometimes used to describe the body of a post mill.

Crosstrees — Horizontal timbers forming the trestle of a post mill, along with the post and quarterbars. They are placed at right-angles to each other; the post is quartered at its lower end so that it can fit over them at their point of interesection.

Cullen Stone — Rarely found type of millstone imported from Cologne (now Koln) in Germany, and used to grind oats and barley.

French Stone — Or French Burr stone; millstone of a smooth texture, quarried in Central France and used in the production of flour.

Flour Machine — Name sometimes given to a dressing machine.

Fulling Mill — Mill (usually a watermill) for the production of cloth.

Hemlaths — Longitudinal members forming the edges of a sail frame.

Hollow Post Mill — Windmill where the machinery is in the lower part of the building and the drive to it from the windshaft and brakewheel, or their equivalent, taken down through a hollow post.

Lucam — Boarded wooden structure on the outside of a watermill building, enclosing a loading door and containing a hoist by which sacks of meal or grain were brought into or out of the mill.

Peak Stone — Millstone of a rough texture, quarried in the Peak District of Derbyshire and used to grind animal feed, which windmills and watermills also produced from time to time in addition to flour.

Quarterbars — Diagonal timbers bracing the post of a post mill to the ends of the crosstrees to complete the trestle structure.

Sheers — Longitudinal timbers forming the sides of the base frame of a windmill cap.

Shot Tower — Building used in the manufacture of lead shot (*ie* cannonballs and the like). The molten lead was dropped inside the building from the top to land in water which froze it into the globular shape it assumed while falling.

Single-sided Sails — Sails with the framework on one side only of the stock. Those with the framework on both sides may be called double-sided. With spring or patent sails one speaks of 'single-shuttered' or 'double-shuttered' sails.

Smutter — Machine for removing impurities (smut) from grain prior to grinding.

Stage — (or gallery) Platform encircling a smock or tower mill to give access to the sails for maintenance.

Stock — The timber to which the whip of a windmill sail is attached.

Uplongs — Longitudinal members of the framework of a common sail.

Weatherbeam — Timber supporting the front end of a windshaft.

Weatherboarding — Boards covering the framework of a windmill structure and the roof of the cap of a smock or tower mill.

Whip — The timber to which the frame of a windmill sail is attached.

Winding — The task of turning the body of a post mill or the cap of a smock or tower mill so that the sails face into the wind.

MILLWRIGHTS KNOWN TO HAVE OPERATED IN MIDDLESEX

(*Kelly* = *Kelly's Directory*)

Alfred Beasley, High Street, Uxbridge (*Kelly* 1867, 1874, 1890, 1899).
Juno Bulcraig, Hillingdon.
Thomas Collard, Yiewsley (*Kelly* 1882), later of Collard, Fraser and Scott, Yiewsley (*Kelly* 1890)
John Cooper, West Drayton.
John Cooper, Old Street, mentioned in *London Gazette* 11-15/12/1798 in which he relinquished business in favour of Thomas Cooper, probably his son.
Cowper of Poplar (see entry on Newgate in Windmill Gazetteer).
Benjamin Dobson, millwright and engineer, High Street, Uxbridge.
Hunter and English, Bow (*Chelmsford Chronicle* March 1838). Also listed in *Kelly's Directory* when their address is given as in the High Street.
William Dobson Gales, Rockingham Ironworks, Uxbridge (*Kelly* 1867, 1874).
Frederick William Jackson, Kingsbury Ironworks, Kingsbury Road, Balls Pond (*Kelly* 1878).
J. Leach & Son, Anchor brick mould works, High Street, Hounslow (*Kelly* 1867).
John Leach, Whitton Road, Hounslow (*South Western Directory* 1865-6). Probably same as J. Leach above.
William Leatry, Old Ford, Bow.
George Lewis, Whitton Road, Hounslow (SWD 1865-5).
Charles Newton and Charles Worssam, Kingsland Road, engineers and millwrights, listed as bankrupt in *London Gazette* 10/12/1844.
John Parish, Goswell Street, London (mentioned in *LG* 16/9/1820 as insolvent debtor).
Edward Sadler, engineer and millwright, Walham Green, Fulham (*Pigot* 1839).
Thomas Savage, West Drayton (*Kelly* 1862, 1867, 1874).
George Stacey & Sons, Uxbridge Ironworks, office 52 High Street, Uxbridge (*Kelly* 1882)
Thomas Taylor, Hanworth (*Kelly* 1862)
Thomas Taylor, New Road, East Bedfont, Feltham (*Kelly* 1890,1899) possibly same as the Thomas Taylor above.
Thomas Taylor, Staines, listed as bankrupt in *London Gazette* 11/12/1824.
Henry Timmins, St John's Works, The Green, Southall (*Kelly* 1874).
John Tucker, Harlington (Robson's *Middlesex*, 1837, and *Pigot* 1838).
J. West, Hillingdon (*Kelly* 1867).
William Winterbourne, South Street, Isleworth (*Kelly* 1874, 1878).

Writing in his book *The London Tradesman* in 1747, C. L. Campbell states of millwrights and millers: 'These two tradesmen are better understood in the country than in the city, though there are some who live in the city and are concerned in mills in the country. The millwright's is an ingenious and laborious business in which there is a great variety, according to the different principles upon which the mill is constructed, but the wages given to journey men is no more than that of a common carpenter. A working miller earns about Ten Shillings a week'.

Bibliography

Buckland, Stephen *Technical Notes On 16th and 17th century London Windmills* Newcomen Society Transactions Volume 60 1988-89.

Crocker, Glenys *The Gunpowder Mills Gazetteer* (Society for the Protection of Ancient Buildings 1988)

Farries, K. G. *Essex Windmills, Millers & Millwrights* (Charles Skilton 1981-8: Vol 1 1981, Vol 2 1982, Vol 3 1984, Vol 4 1985, Vol 5 1988)

Farries, K. G. and Mason, M. T. *Windmills of Surrey & Inner London* (Charles Skilton 1966)

Holt, Richard *The Mills Of Mediaeval England* (1986)

Philo, Phil *Report on the Hounslow Gunpowder Mills* undertaken for Friends Of The Earth in 1984, now held at Twickenham Public Library

Philo, Phil and Mills, John *The Bedfont Gunpowder Mills* "London Archaeologist" Vol 5 No 4 August 1985

Reid, Kenneth *Watermills Of The London Countryside* Vols 1 & 2 (Charles Skilton 1987 and 1988)

Wailes, Rex *The English Windmill* (Routledge & Kegan Paul 1954; reprinted 1971). The standard work on the subject

Wenham, Peter *Watermills* (1987)

Key to Caption Credits

GB	Guy Blythman
BM	British Musuem
SB	Stephen Buckland
OCR	Mrs O. Clive-Ross
CLA	Cricklewood Library & Archives
FHM	Forty Hall Museum
GL, COL	Guildhall Library, Corporation of London
GLRO (M & P)	Greater London Record Office (Maps & Prints)
HAD	Hackney Archives Department
HHS	Hillingdon Heritage Service
HLS, CLLHC	Hounslow Leisure Services: Chiswick Local History Collection
JHA	Julian Harrap Architects
JKM	J. Kenneth Major
LBELS	London Borough of Ealing Library Service
LBOEL	London Borough of Enfield Libraries
LBRUT	London Borough of Richmond-upon-Thames
LBTH, LHLA	London Borough of Tower Hamlets, Local History Library and Archives
MO	Mrs Margery Orton
MOL	Museum of London
RLTMT	River Lea Tidal Mills Trust
UB	*Uxbridge Gazette*
WG, UOM	Whitworth Art Gallery, University of Manchester)

Line Drawings

PB	Paul Brown
SB	Stephen Buckland

INDEX

Figures in *italics* refer to illustrations; **bold** figures indicate main entries.

SUBSCRIBERS

Presentation copies

1 Wind- and Watermills Section, Society for the Protection of Ancient Buildings
2 Greater London Record Office
3 History Department, Southampton University

4 Guy Blythman
5 Clive & Carolyn Birch
6 Mark Blythman
7 Margaret Walls
8 D.T.N. Booth
9 Matthew Licence
10 Miss H. Phippen
11 J. Kenneth Major
12 K.R. Berry
13 Marian Chamberlain
14 R.A. Marsdon
15 Nick Kelly
16 Betty Richardson
17 P.S. Jarvis
18 Tony Shaw
19 Dr Barry Job
20 Michael Yates
21 Martin Watts
22 John Pelling
23 Alan Stoyel
24 Russell Jones
25 Derek Stidder
26 R.W.D. Sharp
27 Tony Bonson
28 Richard A. Bramwell
29 Geoffrey Holman
30 John Vince
31 Gareth N. Hughes
32 Len O'Flanagan
33 D. S. Paterson
35 Enfield Golf Club Ltd
36 Lawrence J. Turner
37 B.P. Jones
38 Donald W. Muggeridge
39 Barnet Libraries
40 Les Osborne
41 R.H. Thompson
42-43 London Borough of Enfield Arts & Libraries
44 Colin Squire
45 John C. Sheaf
46 Philip & Helen Wareham
47 E. Sammes FSA
48 Frank W. Gregory
49 Albert & Patricia F.C. Parr
50 N. J. Pollard
51 Ted Crouchman
52 Niall Roberts
53 Mrs K.M.C Brown
54 B.L. Duttson
55 Stephen Buckland
56 R.V. Clarke
57 P.E. Love
58 Paul A. Brown
59 E.S. Worrall
60 D.J. Hutson
61 Mrs L. Kenny
62 M. Harverson
63-64 Local Heritage Services, Hillingdon
65 National Monuments Records Centre
66 David Stuttard
67 K.F. Kirsopp
68 M.O. Price
69 Mrs E. Coles
70 J.S. Tasker
71 Gunnersbury Park Museum
72 M. Hind
73 Alan & Glenys Crocker
74 J. Harrap
75 R. Hawksley
76 G.M. Saul
77 B.A. Harrison
78 C.K. Bragg
79 R.W. Smith
80 Mrs Elaine Maul
81 R.W. Boustead
82 Norman S. Coles
83 D. Filmer
84 A. Lewis
85 London Borough of Richmond Upon Thames Libraries Dept
86 M. Hoather
87 J.H. Venn
88 James A. Hillier
89 Ken Jones
90 J.C.F. Smith
91 Robert Paterson
92 A.M. Lewis
93 Phil Dunne
94 Susan J. Moore
95 Miss M. Walls
96 V. Wilson
97 Simon Jones
98 Michael Oakley
99 J. Hayes
100 M. Duffau
101 N. Catford
102 C.J. Palmer
103 Mrs M. Darlison
104 Mrs H.J. Hislop
105 Malcolm D. Boyt
106-107 Surrey County Libraries
108 M. Walker
109 M.J. Ireland
110 Chris Lynn
111 Richard Jory
112 Colin Smith
113 David G. Neaves
114 J.W. Hill
115 P.J. Senior
116 Joan G. Lawrence
117 Keith Preston
118 E. Linwood
119 M. Lowndes
120 Dr Barry Job
121 David Wright
122 Commander J. Rayner
123 D.A. Rogers
124 G.T. Riley
125 Brian E. Hill
126 Peter C.J. Dolman
127 J. Kenneth Major
128 S.W. Bartlett
129 P.J.W. Ambler
130 A. W. Sexton
131 Huw Bowles
132 A.C.L. Wall
133 Bob Peterson
134 Matthew Stuart Licence
135 H.M. Hoather
136 K.S. Phoenix
137 E.F. Goatcher
138 Paul Butler
139 N. & M. Bush
140 D.A. Chambers
141 D. Robinson
142 Jo Roberts
143 P. Mills
144 London Borough of Camden
145 Nigel Moon
146 D.R. Honour
147 Sheila Rosamond
148 Mrs E.P. Murray
149 Fred Hamond
150 Sharon Maltby
151 Robert Sier
152 M.J. Goatcher
153 Arthur C. Smith
154 Mrs C. Burden
155 W. Griffiths
156 J.S. Uff

157 Mike Tunnicliffe
158 Dominic Burrows
159 R. Stearn
160 A. Kite
161 Alfred Friday
162 Mr Giles
163 J. Smith
164 L. Szkoda
165 H.M. Hoather
166 Ronald Harry Josey
167 Fred Atkins
168 Alan J. Mitchell
169 Tim Hampton
170 G.S. Boulton
171 Brian J. Lee
172 D. Neaves
173 David Lindsey
174 A.J. Shailes
175 Peter R. Frazer
176 James Butson
177 Michael John Fuller
178 M.H. Edgeworth
179 Douglas White
180 C.M.F. Judge
181 Mrs D. Cole
182 B.K. Tate
183 Institute of Historial Research
184 Josephine E. Duggan
185 R.J. de Little
186 Ton Meesters
187 J.B. Calladine
188 J.B. France
189 Graham Hackney
190 B.J. Gray
191 Malcolm T. Tucker
192 R. Swift
193 D.S. Burrows
194 P.M. Kennedy
195 Morna D. Hooker
196 David Bushnell
197 Iain Cotterill
198 C.H. Bartlett
199 Mrs K.G.A. Bullock
200 Jonn Wright
201 Mrs D. Blythman
202 Guy Blythman
203 Robert Estall
204 Mrs M.L. Plested
205 C.P. Freeman
206 R. Slaughter
207 Robert F. Starling
208 David Barmes
Remaining names unlisted